FIRES OF LIFE

Fires of Life

Endothermy in Birds and Mammals

BARRY GORDON LOVEGROVE

FOREWORD BY ROGER S. SEYMOUR

Yale

UNIVERSITY PRESS

NEW HAVEN AND LONDON

Published with assistance from the foundation established in memory of
Calvin Chapin of the Class of 1788, Yale College.

Yale University Press books may be purchased in quantity for educational, business,
or promotional use. For information, please e-mail sales.press@yale.edu (U.S. office)
or sales@yaleup.co.uk (U.K. office).

Set in Electra LH type by Newgen North America, Austin, Texas.
Printed in the United States of America.

ISBN 978-0-300-22716-1 (hardcover : alk. paper)

Library of Congress Control Number: 2018960623

A catalogue record for this book is available from the British Library.

This paper meets the requirements of ANSI/NISO Z39.48-1992
(Permanence of Paper).

10 9 8 7 6 5 4 3 2 1

For Christopher

CONTENTS

CONTENTS

CONTENTS

You hold in your hand a unique book that I wish I had written. It considers the most important of evolutionary innovations: the appearance of highly active, warm-blooded species some 250–300 million years ago. This development occurred in two lineages of vertebrates, in the mammals and in the archosaurs ("ruling-reptiles") that gave rise to dinosaurs and birds. Because these groups were warm blooded, they have dominated the face of the earth right up to the present day. How these animals became warm blooded and what sort of conditions might have been responsible should interest everyone.

The tree of life has many branches. Most people are interested in the branches closest to human beings. Visitors to a zoo, for example, are generally more attracted to primates than to other vertebrates, and prefer mammals and birds over reptiles, amphibians, and fish. Invertebrate animals generate such little interest that they are hardly represented in zoos at all, despite being more numerous on earth. Part of the attractiveness of mammals and birds is that they are much more active and amusing than other animals. Their high levels of activity are a result of their being *warm blooded*, the common term that describes creatures with a body temperature higher than that of their surroundings. *Cold-blooded* vertebrates, including most living reptiles, amphibians, and fish, have low metabolic rates and body temperatures more or less the same as their environment. Physiologists call the warm-blooded groups *endotherms* (from the Greek words *endon*, meaning within, and *therme*, meaning heat), because a high metabolic rate inside the body is the principal origin of their body heat. The cold-blooded animals are *ectotherms* (Greek *ektos*, meaning outside), because their body heat is primarily determined by the environment around their bodies.

Endotherms and ectotherms are like chalk and cheese. Aside from being able to keep their body temperatures high, endotherms have metabolic rates some ten times greater than ectotherms, and consequently are able to be more active for longer periods and generate more power for subduing prey, and for defense. As the cost for this activity and endurance, they require about ten times as much food as ectotherms. This energy-expensive lifestyle is matched by the ability of endotherms to find food under cool conditions, when ectotherms cannot function. The high metabolic rate of endotherms allows some of them to travel long distances in search of food; ectotherms cannot. When the cost of obtaining food is greater than the energy contained in it, some endotherms, called *heterotherms*, drastically reduce their metabolic rates and enter a temporary state of torpor and hibernation. Conversely, some ectotherms, such as lizards, snakes, extremely active fish, and most flying insects, can increase their metabolic rates high enough to significantly raise their body temperatures.

Evolutionary physiologists such as Barry Lovegrove ask questions about the origin of endothermy among vertebrates. Because it is not possible to take direct measurements of the metabolic rates and body temperatures of extinct animals, we are left to speculate on how and when the transition from ectothermy to endothermy occurred in evolutionary history. These conjectures are based on the patterns of endothermy in living mammals and birds, as well as the form and function of fossil species.

The evolution of endothermy has been the subject of many papers over the years, but this is the first book I know of that deals directly with mammals and birds on their own. The first chapters of the book consider the fossil evidence and the lineages of birds and mammals, giving an idea of the circumstances and environments that may have played a part in the appearance of endothermy. The middle chapters are case studies of extant mammals that offer insight into the kind of early mammal that Barry envisions. These chapters record highly personal research experiences, most interestingly in hot and humid Madagascar, where tenrecs apparently do not require endothermy. The final chapters return to the physiology of endothermy and temperature regulation, and include several hypotheses for single causes of the evolution of endothermy.

Barry Lovegrove is an expert on the subject of endothermy in living species, and he lives near one of the most prolific sources of early mammal fossils. I have followed Barry's research for decades, from his initial work on subterranean mole rats from southern Africa. These social animals form colonies and feed entirely underground, burrowing blindly through the soil until they encounter plant tubers that they strategically use for food. Barry measured the energy cost of burrowing and compared it with the energy gain from food, yielding one of the most elegant studies in energy economics in the ecological literature. I have presented this material as a model study in my course on environmental physiology for the past thirty years, because nothing beats it. Among Barry's other notable contributions, he introduced the concept that the variability in the metabolic rates of different species of mammals depends on body size, being most variable in the smallest and largest animals, and less so in the middle range of size. In the year 2000, he called this the bow-tie pattern, because the points on a graph were constrained at body sizes of about 350 grams, but flared out on either side, just like a bow tie. The name stuck. Craig White and I confirmed this pattern in 2004 through an analysis of more than six hundred species of mammals, although the reason for it remains elusive.

The effect of body size is apparent in every aspect of animal biology, not only physiology, but also anatomy, ecology, and behavior, and the effects are usually not proportional to body size. This book recognizes that the way ecophysiologists study the effect of body size is by plotting data on graphs with logarithmic axes, setting equations and statistics to the patterns, and then trying to decipher their meanings. One of the first physiologists to do this was Max Kleiber, who wrote a book in 1961 called *The Fire of Life: An Introduction to Animal Energetics*, which has been a touchstone for physiologists ever since. The title of the book you hold is an homage to Kleiber, but its content mercifully spares you the tedious analytical detail of allometric equations. Rather, Barry concentrates on the phenomenon of endothermy in relation to body size in a general sense.

This book is highly original, especially as it mixes hard science with tales of the field research that gave rise to it. It is written in a conversational and very accessible style. Those who want substantiation and further information on the subjects are rewarded with a list of references at the end of the book. There is a treasure trove of information in this volume. I found myself highlighting many passages that contain insights and ideas that I did not appreciate until reading this book. Many of the references are quite recent. That said, the book is not meant to be a review of all of the literature. It is refreshing that Barry mentions some studies that have evidence contrary to his views, and he says explicitly that he is capable of changing his mind, a characteristic of a true scientist.

Barry Lovegrove has been a leader in the environmental physiology of mammals and birds for more than three decades. He focuses on the energy, water, and thermal exchanges between these animals and their environments, often attracted to animals with extreme behaviors and physiologies that are adapted to them. His early work on subterranean social mole rats, for example, showed variable metabolic rates and loose temperature regulation in response to extremely energy-consuming burrowing for food in one of the least productive and unpredictable environments of the world, the Kalahari Desert. This led to a more general interest in the biogeography of energy balance and temperature regulation in several species that enter states of dormancy to conserve energy.

Many environmental physiologists study single species that live in extreme environments, described by one of the founding fathers of environmental physiology, George Bartholomew, as "Oh my!" animals. Such studies are fascinating in themselves (prompting the exclamation), and are important because they help to define the limits of adaptation, but they sometimes do not lead to a broader biological significance. Not content with limiting himself to Oh my! studies, Barry has synthesized his results and current literature into a series of general publications on the evolution of present patterns of adaptation and strategies for survival. Often single-authored in prestigious journals, these synthetic works consider broad sub-jects, including mammals that may discontinue temperature regulation for hours or months, the role of sociality among wholly subterranean mam-mals, the relationship between basal metabolic rate and mammal distribu-tions, the patterns of temperature regulation in relation to season, and even an explanation of why most placental mammals have a scrotum. With his former student Andrew McKechnie, Barry has also written a general review of heterothermy in birds that has been very influential. Recently he has become interested in the implications of his research for the evolution of endothermy, resulting in two significant reviews that are the precursors of this book.

Whether you are an interested layperson or a professional biologist, I am certain you will learn a great deal from this book, as well as enjoying the author's entertaining stories about his research in the quest to understand the evolution of endothermy.

Roger S. Seymour
Professor Emeritus, University of Adelaide, Australia

*F*ires of Life is a scientific story that will help you to understand why your blood is routinely warmer than that of a gecko. The book is based on a career of research on the energetics of mammals and birds and on ideas generated by living and working in the southern hemisphere, far away from mainstream thinking in contemporary evolutionary physiology; it is inspired by an uneasiness that some ideas that science might have found useful in the eternal human quest to understand our universe might never have been debated or tested.

I am an evolutionary physiologist. The quest of all evolutionary physiologists is to understand the diversity of physiological patterns that make animals work in different ways in different environments. We ask questions such as, Why are some animals hotter than others? Why are some bigger than others? Why do some have saltier blood than others?

Why do birds and mammals have warm blood routinely, whereas other animals do not? This is the foundational question that I ask and attempt to answer in this book. It is by no means a trivial question. It is perhaps one of the most important questions in evolutionary biology. The answers will palpably boost your understanding of yourself and supply insight into the stupendous diversity of warmth in animals of the present and the past, including the dinosaurs.

Birds and mammals are the only animals that have the physiological machinery to keep themselves warm throughout a very cold day. They are endotherms, which means that they can generate heat on demand in their own bodies to balance heat that is lost from their bodies. Ectotherms are animals that do not, as a rule, have the capacity to produce heat on demand to replace lost heat. We once knew ectotherms as "cold-blooded" animals and endotherms as "warm-blooded" animals. For physiologists, however, these are inappropriate terms, because some lizards can attain warmer blood than that of mammals by basking in the sun, and some mammals can maintain a subzero body temperature when they are hibernating.

The massive difference between ectothermy and endothermy is the costs that are involved: endothermy is about ten times more "expensive" than ectothermy. So why did endothermy evolve when it is such an expensive way of living? Endothermy must have benefits that outweigh the costs—otherwise it would never have evolved. All evolutionary processes are driven by relative costs and benefits—what biologists term trade-offs. But what are the benefits of being consistently warm? How do these cold-hot trade-offs work?

Endothermy is indeed one of the greatest enigmas of evolution. The understanding of the evolution of endothermy has a very punctuated history. In the 1970s a flurry of hypotheses were published by pioneering thinkers such as Robert Bakker, Alfred Crompton, Brian McNab, Albert Bennett, John Ruben, and Roger S. Seymour on why or how endothermy evolved. These early ideas are all stitched into my timeline of the 300-million-year journey toward the supraendothermy that we see today in some birds and mammals. These early explanatory models, as well as several that Colleen Farmer and Pawel Koteja published three decades later, were "single-cause" models, which I argue are insufficient individually to formulate a holistic model on the evolution of endothermy. So the endothermy story has stagnated despite an enormous amount of research—for example, into one of its principal models, the Aerobic Capacity Model. It is hard to understand such stasis given that the evolution of endothermy was possibly the most critical innovation and evolutionary development during the evolution of birds and mammals. The endothermy story has become bogged down.

It is therefore timely to revisit the topic from new and different angles and along expanded time scales. With the insight provided by new paleontological data, such as splendid fossils of Jurassic and Cretaceous birds and mammals from China, for example, it is possible now to link the present with the distant past in a phenological framework—a working time model—on the evolution of endothermy. I truly intend the model to be a source of novel, testable hypotheses rather than a definitive treatise.

The book is not a maverick Gondwana alternative to the status quo; it shares with scientists and the informed reader a multitude of novel, peer-reviewed hypotheses about how, why, and when endothermy evolved in the two great lineages of vertebrates that produced the endotherms: the mammals of the synapsid lineage, and the dinosaurs and birds of the sauropsid lineage.

Part 1 of this book relates the story mostly from what we can glean from the study of fossils and starts more than 300 million years ago, when the first ancestors of the amphibians hauled out onto the muddy banks of swamps and started to eat plants. In Part 2, I tell the story from data that have been collected from living animals, much of it from my own work. These are new data that my students and I have collected in South Africa and Madagascar that support the argument that endothermy evolved for the most part in tropical Permian, Triassic, Jurassic, and Cretaceous forests, in an iterative, pulselike way. These eras were interspersed with intensely hot, arid periods that propelled the transition from ectothermy to endothermy in certain lineages.

I can list on the fingers of my hands the number of evolutionary physiologists who live and work in the southern hemisphere, compared with the many hundreds in the north. The small southern group includes a few who are as fascinated as I am by the evolution of endothermy. We share many common ideas that, in many respects, provide alternatives to those that have been taught and published by northern hemisphere biologists over the past century. I trust that my model and its hypotheses do their passion justice.

Although this book undoubtedly presents a few novel ideas that have yet to be vetted by my peers, for the most part it is centered on a peer-reviewed monograph published in *Biological Reviews*. The book itself also underwent several review processes managed by Yale University Press. For the sake of continuity of my story, I needed at times to take sides in some of the ongoing quarrels that may or may not have a bearing on the story of the evolution of endothermy. The debate about whether the major bird and mammal radiations occurred before or after the Cretaceous-Cenozoic boundary is especially vehement.

There are two types of physiology: mechanistic physiology, which asks how, and evolutionary physiology, my favorite, which asks why. Evolutionary physiology makes no sense, of course, without an appreciation of mechanistic physiology; but I have tried to minimize discussion of mechanistic physiology. Where it is essential for a continuity of understanding, I have relegated it to the appendixes. Students of physiology can simply refer to the multitude of physiology textbooks that explain how animals work. This book focuses on evolutionary physiology, on why there are differences in how animals work.

I cover climate change over 300 million years and trawl through the earliest proto-mammals and proto-birds that persisted through the Triassic heat. I discuss their remarkable explosion of diversity in the Jurassic, with animals that were the ancestors of the pterosaurs, dinosaurs, birds, and mammals.

It is a hot story.

ACKNOWLEDGMENTS

*F*ires of Life has drawn so heavily on my work and that of my students over the past three decades that I cannot ever sufficiently acknowledge all who have had an influence on my thinking and research. How much of what I write, for example, can I attribute to the author of my university undergraduate animal physiology textbook, the wonderful animal physiologist Knut Schmidt-Nielsen? One thing is clear in my mind, though: I could not have written *Fires of Life* were it not for my postgraduate students Kerileigh Lobban and Danielle Levesque, whose hard work in Madagascar provided some of the novel data and certainly the inspiration, circumstance, and conceptualization for the book. They afforded me my "penny-dropping" moment in Madagascar. I am particularly grateful to Keri and Danielle for persisting through a nasty Madagascan coup d'état on an ever dwindling research budget.

The title *Fires of Life* is derived, of course, from Max Kleiber's classic work *The Fire of Life: An Introduction to Animal Energetics*. I am grateful to Karen McGregor and William Saunderson-Meyer for their encouragement during the writing and publishing process and, most importantly, for insisting that I choose my highly desired Kleiberian title for the book.

Several colleagues were kind enough to review draft chapters at my request: Kerileigh Lobban, Danielle Levesque, Andrew McKechnie, Ben Smit, and Bruce Rubidge. Mark Brigham and Roger S. Seymour provided comments on chapter drafts at the request of Yale University Press. I am grateful for their positive feedback.

It was a pleasure to interact with Yale's acquisitions editor Jean Thomson Black, and her assistant, Michael Deneen, who helped me to improve my initial book proposal. I am grateful to Marilyn Martin, who did a marvelous job editing the manuscript, and to Phillip King for overseeing the production process.

I started to write *Fires of Life* while on sabbatical in 2015. Several colleagues stood in for me in my regular academic duties and teaching program, which added greatly to their workload. In particular, I thank Benny Bytebier.

FIRES OF LIFE

PART ONE

The Story in the Rocks

ONE

A Narrative of Travels

Charles Darwin did not embark upon the HMS *Beagle* to sail around the world for five years with the outward intention of gathering information to enable him to write *The Origin of Species*. Darwin's counterpart, Alfred Wallace, also did not embark upon his remarkable travels through the islands of Southeast Asia to form ideas and pen stories for his book *The Malay Archipelago*. Both of these classic books on evolution were conceived *after* the respective voyages by their authors and proffer, today—unarguably for me—the greatest idea ever offered by humans: a theory of evolution.

In his opening chapter, Wallace wrote: "It may be though that the facts and generalizations here given, would have been more appropriately placed at the end rather than at the beginning of a narrative of the travels which supplied facts. In some cases this might be so, but I have found it impossible to give such an account as I desire of the natural history of the numerous islands and groups of islands in the Archipelago, without constant reference to these generalizations which add so much to their interest."[1]

When I started my first real job, I did not choose a research topic and pin a sticky note to my computer screen that read "Research topic: The evolution of endothermy." Not at all. The topic crystallized while I was sitting beside a smelly, steamy swamp in Madagascar one very hot afternoon. It happened after I reached a threshold of ideas that suddenly, upon a trigger, poured into the mold that is *Fires of Life*. So I, too, will mix up my timeline of thoughts and experiences to formulate a sensible story about why birds and mammals are endotherms: why they have warm blood, as a rule. I want to share with you a narrative of my travels, not only from here to there but also from then to now. Like Wallace, I really did not know where my travels would take me initially or that I would hoard for so long all of the bits and pieces of a puzzle that one day popped into place beside a swamp to allow me to realize the story that is *Fires of Life*.

Evolutionary physiology alone cannot provide all the information that we need to answer questions about the evolution of endothermy. I have leaned heavily on the work of paleontologists, climatologists, geologists, ecologists, behaviorists, anatomists, and mathematicians to mention the obvious, to try to gain a holistic vision of the evolution of endothermy. This huge diversity of scientific disciplines is the reason that no book similar to *Fires of Life* has ever been attempted. The author would have needed to be a scientific jack-of-all-trades to accomplish this, which is an insane expectation. I have given it a go not because I am insane, I think, but merely because I happened to stumble upon a *framework* to follow, a framework that popped into my head beside that stinky swamp in Madagascar. All I needed to do then was look in the right places for the right information, at the right time, and make sense of it all. *Fires of Life* is a working model.

Disagreement among scientists on certain topics is the propellant of science because it sustains the alternatives that extricate it from religion. Science proceeds from one fashionable, testable guess to another, which is what makes it the most progressive tool of knowledge acquisition that we have. At the time I wrote this book in the summer of 2016, I could at best provide a snapshot in time of our collective thinking. I believe that we have accumulated sufficient knowledge on the topic now to have a good stab at answering the pillar question and to put together a sensible and, at the same time, I think, fascinating story.

Let me start the story by talking a bit about heat and energy to pose a background of endothermy. I live in Pietermaritzburg, a town in South Africa that the novelist Tom Sharpe described as "half the size of a New York cemetery and twice as dead." It is in the province of KwaZulu-Natal and was named after two Voortrekker leaders, Pieter Maurits Retief and Gert Maritz. The town was founded in 1838 following the Battle of Blood River, where twenty to thirty thousand Zulus under King Dingane fought 470 Voortrekkers (South African pioneers) led by Andries Pretorius—spears against muskets. About three thousand Zulus were killed and three Voortrekkers wounded. King Dingane murdered Retief before the great battle, though, along with about a hundred companions, after the signing of a deed of concession of land at Dingane's kraal UmGungundlovu near present-day Eshowe. The remains of Retief's party were left lying in the open on a ridge called Hlomo Amabuto for the hyenas to eat and the flies to breed. Andries Pretorius recovered the signed concession from Retief's leather pouch after he led a victory commando to UmGungundlovu four days after the Battle of Blood River.

After the drama, the Voortrekkers chose the location of Pietermaritzburg in the Duzi River valley wisely; it is quite cozy in winter, compared with the surrounding midlands, the foothills of the Drakensberg Mountains. We don't bother with central heating during the winter: it is not consistently cold enough to warrant the cost of double-glazed windows or a heating system. There are no fitted carpets in my house. The floors are wood or cool clay tiles with Persian carpets flung over them. In winter we haul out the thicker duvets and huddle around an occasional log fire. And so, too, do our dachshunds. It is their second-best thing, taking long naps in front of the fire during the winter off-season following the exhausting summer gecko-hunting season, their best thing. Geckos cannot be hunted in winter because they do not move around. They are too cold. They are ectotherms. They rely exclusively on a warm air temperature to warm up.

In Europe and North America central heating is commonplace, and most people can afford it. Heat is measurably expensive. It is a form of energy that can be transferred from one body to another through kinetic contagiousness. It flows downhill, always, from a high to a low temperature, from a warm room to a cold room, from warm skin to cold air. Under normal, healthy conditions, our body temperatures are regulated slightly above $37°C$ ($98.6°F$). If we are exposed to an air temperature lower than this, heat will be lost from our bodies and our body temperature will start to decrease.

5

Chilled blood is not good for the brain and the heart, especially in humans. Those organs start to malfunction. The brain goes fuzzy, and the heart fibrillates. Pathological hypothermia—uncontrolled below-normal body temperatures—is deadly in animals that have evolved to have a constant body temperature. In mammals and birds, nonpathological hypothermia occurs in hibernators only.

Early humans did not have central heating. They used fires to heat their caves and huts. In theory, though, these early humans did not actually need a fire to keep a constant body temperature. Humans learned how to change their insulation, as circumstances demanded, by adding or removing from their bodies the furs of other animals. One obvious advantage to having this fur-swapping capacity was that humans did not need to molt between seasons, which is an expensive process and can render animals vulnerable to extreme, unseasonal weather events. Most mammals and birds that live in the cold higher latitudes molt seasonally. As summer approached, humans merely peeled off the fur layers and by midsummer pranced around in loinleathers, as the San still do to this day in the Kalahari Desert, bare-bummed. Indeed, a naked skin and the ability to sweat freely allowed early humans on the African plains to hunt their antelope prey by running them down to heat exhaustion in the hot part of the day. Evidence of the long-term selection for this ability is commonly displayed to this day in the supreme domination by East Africans of long-distance running races.

Endothermy is defined as the capacity to produce heat on demand from within an animal. In birds and mammals, this heat is generated by the internal organs. To keep the body temperature constant, heat that is lost must be replaced by heat generated from inside the body, and this capability is known as endothermy (from Greek: *endon*, "within"; *therme*, "heat"). When the body temperature of the animal is higher than the air temperature or the temperature of surrounding objects, heat leaves the body through the physical processes of radiation, convection, and conduction. All mammals are endotherms, apart from a few screwball exceptions we will meet in later chapters, and so, too, are birds. Endothermic characteristics occur in some invertebrates—insects—but that is another story for another book. Endotherms produce heat using their own metabolic machinery from the foods that they eat: carbohydrates, proteins, and fats, with some smart little tricks used by their cell membranes. Geckos cannot do this.

You can adjust the temperature in your water heater, what we call a geyser in South Africa, by twiddling the thermostat setting. If the temperature in the heater falls below your chosen setting, the heating element switches on and heats the water. As the temperature increases past your setting, the element switches off and the water starts to cool again. A recorded time tracing of your heater temperature would look like a snaky wave centered on your setting. The endotherm's body temperature is also regulated around a set-point body temperature in a way similar to that of your water heater. The mechanism of regulation between species varies little, but the settings vary a lot. The temperature is regulated by the hypothalamus in the brain from sensory temperature information fed to it from the skin and other parts of the body. The actual setting is pretty much a hard-wired feature of each species, although it can be adjusted on a daily and seasonal basis or when an animal is starved.

The minimum amount of energy that animals need to stay alive is provided by what is called the basal metabolic rate (described in Appendix 1), and the heat that is generated comes from the internal organs: the heart, kidneys, liver, and intestines. Heat does not come from muscles if the animal is at rest and no extra heat is needed to stay warm. The mitochondrial membranes in the internal organ tissues are leaky; Na^+ and K^+ ions and protons cross the membranes in a constant, uncontrolled way. The constant leakiness produces a proportional amount of heat. The internal organs of endotherms also have more than double the mitochondrial membrane volume and a higher mitochondrial density than those of ectotherms. That is why, at rest and gram for gram, an ectotherm has a resting metabolic rate about ten times lower than that of an endotherm.

Birds and mammals can regulate and balance the amount of heat that is produced against that which is lost to control a fairly constant body temperature. Body temperature, at any point in time in an endotherm's life, is a product of all of the thermoregulatory and endothermic processes and interactions in the body. Body temperature, therefore, cannot substitute as a direct measure of endothermy. But because body temperature is easy to measure, I have employed it in this book as a proxy or measure of the level of endothermy in birds and mammals to make discussion easier. They are not exactly the same thing, but temperature can serve as a proxy because there is a positive relationship between an insulated endotherm's capacity for heat production and its body temperature.

This might not have been the case, though, in the earliest stages of endothermy, when insulation could have been lacking in ancestral synapsids and sauropsids. On the basis of simple statistics (percentiles of frequency distributions), the hottest 20 percent of endothermic mammals I have called "supraendotherms"; their body temperatures are higher than 37.9°C (100.2°F).[2] The next group, the "mesoendotherms," form the majority of the mammals (60 percent); their body temperatures fall between 35°C (95°F) and 37.9°C (100.2°F). The coolest 20 percent of endothermic mammals are the "basoendotherms"; their body temperatures are lower than 35°C (95°F). We can use the same procedure for the body temperature data available for birds;[3] supraendothermic birds are warmer than 42.5°C (108.5°F), mesoendothermic birds have body temperatures between 40.4°C (104.9°F) and 42.5°C (108.5°F), and basoendothermic birds are cooler than 40.4°C (104.9°F). These three levels of endothermy make it easier to categorize individuals or groups of birds and mammals that share a similar endothermic status. They also assist us in understanding how and where changes in endothermy occurred over the past 300 million years. Moreover, a particular endothermic status can be pinned to particular forms and functions of certain groups of birds and mammals. For example, as I argue later, all mammals that run fast appear to be supraendotherms.

It is evident from these endothermy categories that the body temperature set point varies between species of birds and mammals by a surprisingly large amount, around 15°C (27°F). The highest regulated body temperature ever measured in an endotherm was 44.6°C (112.3°F), in the somber hummingbird (*Aphantochroa cirrochloris*) from Brazil, the avian supraendotherm par excellence.[3] Gram for gram, birds generally have higher body temperatures than mammals. The coolest endotherms are mammals that live underground, such as the naked mole rat (*Heterocephalus glaber*) and the marsupial mole (*Notoryctes caurinus*), which have normal body temperatures of around 30°C (86°F). These are the basoendotherms par excellence. The underground realm is buffered from the daily temperature cycles aboveground, thus reducing the requirement for as much internal heat production: one explanation for the lower body temperature of some subterranean mammals. The egg-laying monotremes, the echidnas and the platypus, are also typical basoendotherms.

Routinely, the hottest mammals are ungulates, such as the pronghorn and the springbok; lagomorphs, such as hares; or carnivores, such as the cheetah. These are the mammalian supraendotherms. At around 37°C (98.6°F), humans and most primates are mesoendotherms, sandwiched between the cool basoendotherms and the hot supraendotherms.

The mammal that is unquestionably the supraendotherm par excellence—for me, anyway—is the pronghorn (*Antilocapra americana*). This remarkable ungulate has attained the pinnacle of mammalian endothermy. Not only is it the hottest ungulate, apart from a few domesticated animals that don't count in my mind; it is also the fastest. Indeed, apart from the cheetah, which can attain the fastest absolute running speed of all mammals, but only for a few minutes, the pronghorn can do almost the same for hours on end. Pronghorns can cover 11 km (6.8 mi) in 10 minutes, at an average speed of 65 km/h (40 mi/h). No other mammal can come close to matching this.

How the pronghorn got to be so hot and fast is the story of this book. It is a story that starts at the end of the Devonian period, about 359 million years ago, passes through the Carboniferous, Permian, Triassic, Jurassic, and Cretaceous periods, and ends up in the Cenozoic era, the most recent 66 million years—the age of the mammals. It took more than 300 million years of evolution along the line leading to the mammals to reach the endothermic pinnacle we call the pronghorn. Originally, all four-legged vertebrates, the tetrapods, were ectothermic. In the Devonian period these included the ancestors of the amphibians (stem amphibians, defined later) only, because the reptiles, snakes, and lizards had yet to make their appearance on earth. So the story starts cold, but then splits into two stories that lead ultimately to the warmness we see in birds, in the feathered story, and in mammals, in the furry story. These were the success stories of the long journey to endothermy and its persistence in our modern world.

Heat is expensive and is correlated directly with the cost of living. Endotherms need ten times more food than ectotherms, and it needs to be acquired and processed ten times faster. The rationale of life is to breed, to pass on genes to future generations. So even within endotherms we must ask: why should a naked mole rat be able to achieve life's rationale at a fraction of the cost of that of the somber hummingbird? Indeed, why do both of these animals need to maintain body temperatures above 30°C (86°F) to live and procreate? Ectotherms, such as fish, amphibians, and reptiles, can achieve the rationale at body temperatures lower than 30°C (86°F), often substantially lower. The tortoise can theoretically produce as many successful offspring per year as its fabled arch-rival, the hare, but at a much lower cost.

I need a slight digression here to clarify exactly what I mean by costs and benefits.

Energy can be measured in joules, and the rate at which it is used, in watts. In this sense, we can talk of energy costs, such as how much electricity you use in your house. But this is not the same cost as those associated with the rationale of life. Here we are talking about *fitness* costs, and fitness can have benefits, too. Moreover, by fitness I do not mean that one animal has trained more and is therefore "physically fitter" than its rival. The hare does not run faster than the tortoise because it is fitter. In an evolutionary sense, fitness is a measure of how many genes an animal successfully transfers to future generations. If something happens to an animal such that this potential is compromised, it is considered a fitness cost. If something favors the potential, it is a fitness benefit. In an ideal world, which does not exist, adaptation occurs in animals when fitness benefits outweigh fitness costs. We are still trying to figure out exactly how it works.

It is not my objective to defend the theory of evolution in this book. Many authors have done this much more eloquently than I could ever do, so please refer to them if you feel uncomfortable when I talk about the costs and benefits of fitness. Daniel Dennett and Richard Dawkins, for example, are really good at explaining evolution.

It is also not my intention to try to convert you from whatever you think you are into whatever you think I want you to be. Stay with yourself. If you persist with me, though, open up some space in your neocortex to ponder new likelihoods. Park your old neocortical hard drive and all its childhood memories and maps for instant recall when needed, but please format a new one to store novelties for pondering. Ponder, for example, the magnificence of the pronghorn, its speed, grace, and warmth.

I love the concept of mammalness. I do not necessarily love humans, apart from a few, but I do love normal mammals. A normal mammal to me is one that did not acquire, assimilate, and pass on its knowledge, one that did not acquire the ability to speak a language. Language is what makes us human, and it is the only thing that separates us from other mammals. We can pass on what we have learned and experienced in life to the next generation. We can pass on our neocortical maps. In so doing we can exponentially increase our knowledge of how the universe works, and destroy it at an exponential rate as well. If humans had used the full power of their language capacity over the centuries to learn how to live in a sustainable way, our world would be a very different place today. In would be cooler, more peaceful, more biologically diverse, and in better ecological shape.

Before I proceed to the next chapter, I need to introduce you to the main graphic in this book, which shows how I think the evolution of endothermy fits into those parts of the geologic time scale that are of interest to us (Figure 1.1).

I realize it is hard to remember geological names and the time periods they span, but there are not that many, and it is worth getting to know them as a way of grasping the bigger picture. The time scales that concern us are the eras, periods, and epochs. The oldest era discussed is the Paleozoic (551–251.9 million years ago), which is subdivided into six periods, three of which concern us: the Devonian (419.2–358.9 million years ago), the Carboniferous (358.9–298.9 million years ago), and the Permian (298.9–251.9 million

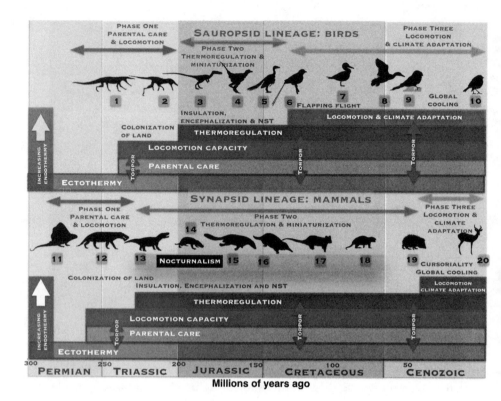

Figure 1.1. A Triphasic Model of the evolution of endothermy in birds and mammals, modified from Lovegrove, B. G., "A Phenology of the Evolution of Endothermy in Birds and Mammals," *Biological Reviews* 92 (2017): 1213–1240. Numbered animal silhouettes (not to scale): 1, *Chanaresuchus*, archosauriform; 2, *Scleromochlus*, early bipedal non-avian archosaur; 3, *Coelophysis*, basal theropod; 4, *Aurornis xui*, close avialan relative; 5, *Archaeopteryx*, basal avialan; 6, *Protopteryx*, basal in Enantiornithes; 7, *Ichthyornis*, basal in Ornithurae; 8, *Anas*, Anseriformes, one of the basal crown bird orders; 9, *Macrodipteryx*, Caprimulgidae, an avian hibernator; 10, *Cossypha*, a passerine bird; 11, *Dimetrodon*, a stem synapsid ("pelycosaur"); 12, *Gorgonops*, a carnivorous therapsid; 13, *Cynognathus*, a cynodont; 14, *Morganucodon*, a basal mammaliaform; 15, *Castorocauda*, a derived mammaliaform; 16, *Steropodon*, a Monotremata; 17, *Pseudocheirus*, a Marsupialia; 18, *Meniscoessus*, a Multituberculata; 19, *Tenrec*, a hibernating afrotherian; 20, *Antidorcas*, a highly cursorial placental mammal. (Animal silhouettes obtained from phylopic.org; figure under a Creative Common license; John Wiley and Sons, License no. 3958130291308)

years ago). The Paleozoic epochs are not important in this story. The next great era was the Mesozoic (251.9–66 million years ago), where all of the big action takes place, and it is divided into three periods: Triassic (251.9–201.3 million years ago), Jurassic (201.3–145 million years ago), and Cretaceous (145–66 million years ago). In the Mesozoic the epochs do concern us because each of the three periods is split into Early, Middle, and Late epochs. So the Middle Jurassic, for example, extended from about 174.1 to 163.5 million years ago. The most recent era is the Cenozoic, known as the age of the mammals, and it covers the past 66 million years. The Cenozoic is subdivided into seven epochs: Paleocene, Eocene, Oligocene, Miocene, Pliocene, Pleistocene, and Holocene. These epochs are important to the story because they involved profound changes in the temperature of the earth and hence influenced the evolution of endothermy dramatically.

I will not fully explain the patterns of the evolution of endothermy at this point, because I will refer to the progression as we proceed through each chapter. Briefly, though, the model identifies three common phases in the phenology (timeline) of the evolution of endothermy. Within each phase the horizontal shaded bars indicate the approximate time of origin of endothermic pulses—increases in basal metabolic rate and body temperature—and the respective proposed driving forces. Phase One is associated with parental care and the aerobic capacity for activity associated with the colonization of dry land by stem synapsids and therapsids in the mammal lineage and by archosauriforms, archosaurs, and nonavian dinosaurs in the bird lineage. Phase Two involves the thermoregulation associated with body size miniaturization, body insulation, increased brain size (encephalization), ecomorphological diversification, and nocturnalism in mammaliaforms (mammals) and theropod nonavian dinosaurs. Phase Three implicates locomotion in the form of cursorial digitigrady (running on toes) and unguligrady (running on tiptoes) in mammals and muscle-powered flapping flight in birds, as well as the climate adaptation in both birds and mammals associated with Cenozoic cooling and vegetation changes. Within each phase, the double-headed vertical arrows indicate the shuttle between endothermy and ectothermic-like states seen today in hibernation and daily torpor, which could have occurred throughout the mammal and bird lineages wherever endothermy had developed. The graphic identifies the geologic times when the great pulses in endothermy probably occurred. Invariably, these pulses were driven by dramatic climate-change events, such as what occurred at the time of the greatest extinction event ever, at the end of the Permian.

Throughout the book I will introduce you to many extinct and living endotherms; stem amphibians, stem synapsids, therapsids, stem mammals, true mammals, stem archosaurs, pterosaurs, dinosaurs, and birds. Each of them lived at some time during the 300-million-year journey to endothermy. Don't despair about their strange names and antiquity. To help you appreciate when they existed, when they went extinct, whether they still exist, and which other species they lived alongside, I have created five "trees of life," what biologists call phylogenies, that show every genus mentioned in the book; these diagrams appear in Appendix 3 as Figures A.1–A.5. Every organism on earth has a unique species name, but many species share a common genus name, because they are genetically more similar to one another than they are to all other species in all other genera. I opted for the generic level in my trees of life because many fossil archaic birds and mammals have yet to be identified down to the species level. Nevertheless, where the species name is known, I have employed it.

Some clarity is needed with regard to terms related to these phylogenies. A simple phylogeny is one in which the "tree" of the phylogeny has a solid stem with no side branches and a canopy of many smaller branches making up the crown of the tree with living species on each stem. Those species at the top of the tree and within the "canopy" above the stem are known as *crown* species. They need not all be currently alive—some may be extinct— just as the crown of a tree could harbor some dead branches. Nevertheless, these extinct species remain within the crown group. All species in such a crown group form a *monophyletic clade,* such as the clade of "Mammals" shown in Figure A.3. The clade is monophyletic because members of the clade evolved from a *single* common ancestor. In a more complicated phylogeny, in which the true arrangement of species has yet to be fully understood (unresolved phylogeny), side branches may occur on the stem of the phylogeny. Again, using the synapsids as an example, these are known as the *stem* mammals, and they would include the "pelycosaurs," which are also known as stem synapsids because they occur at the *base* of the stem, that is, at the origin of the synapsids (as shown in Figure A.1). Because a single common ancestor of stem synapsids cannot be identified as yet, they are known as a paraphyletic group. However, unlike the stem synapsids, the therapsids, which are also stem mammals, do form a well-defined monophyletic clade, and it is for this reason that they were assigned the name Therapsida by Robert Broom, whom we will meet in later chapters. In their case, one branch in the crown gave rise to the stem of another "higher" clade in the tree—the stem that emerged from the crown of Therapsida to give rise to the "Mammalia" (shown in Figure A.2). The same conventions apply to the sauropsid lineage, in which the immediate ancestors of the Archosauria (crocodiles, pterosaurs, dinosaurs, and modern birds), species such as *Euparkeria* and *Garjainia*, are known as stem archosaurs.

Pronghorns inhabit the arid deserts of North America. But let me take you to other places that I know better, to some of the corners of ancient Gondwana where lie some of biology's greatest secrets, hidden for so long from the eyes and minds of the northern hemisphere. I invite you now to share in the narrative of my travels through vast reaches of time and space, from north to south and east to west, in my quest to articulate a story of science. It is a story that will help you better understand how our universe might work, and why you are warm routinely, why a gecko is not, and why the pronghorn and the hummingbird are the luminaries of the chronicle.

TWO

Conquering Land

Ticking Boxes

The first step toward endothermy in the mammal and the bird lineages occurred during the Carboniferous period, when their amphibian-like ancestors wriggled their way out of the swamps and onto mud banks. So the Carboniferous period is the starting point on our long evolutionary journey toward endothermy for a very good reason; it marks the point in time when the two great vertebrate lineages, the synapsids and the sauropsids, parted ways from a common ancestor (as shown in Figure A.1). The synapsid lineage led to the mammals, the furry lineage. The sauropsid lineage led to the reptiles, dinosaurs, and birds, the scaly and feathered lineages. The split occurred about 315 million years ago, so endothermy evolved independently in these two lineages over a very long period of time, from a single ancestor that is generally assumed to have been ectothermic. The innovations and adaptations that evolved to make the haul-out from the swamps successful are some of the fundamental features necessary in the drive toward endothermy and sustained body warmth. Certain boxes needed to be ticked before endothermy could evolve.

The Carboniferous spanned the geological period from the end of the Devonian, 372 million years ago, to the start of the Permian, 299 million years ago. The name *Carboniferous* means "coal-bearing" and is derived from the Latin words *carbo* ("coal") and *fero* ("I bear, I carry"). During this time the world's coal beds were formed. Carboniferous coal captured the ancient air's carbon dioxide when it was fixed in plants by ancient sunlight.

When you burn coal today, in the flames you can see and feel the sunlight of the Carboniferous. You can feel the warmth of an ancient world. You are seeing the sun that shone on the last common ancestor of the synapsids and sauropsids, your presumed cold, ectothermic ancestor.

At the start of the Carboniferous there were no reptiles, mammals, or birds. There were insects and fish and a multitude of amphibious four-limbed animals that dominated the planet. Sandwiched between the lobe-finned fish and the first true reptiles and archaic mammals, four-limbed animals made a lunge, perhaps not the first, at colonizing dry land, and they go by a confusing variety of names. Some biologists call them Tetrapodomorpha (stem tetrapods), others call them Labyrinthodontia, and still others refer to them simply as amphibians or stem amphibians. I will call them stem amphibians because it differentiates them from modern amphibians, and they were, for the most part, dependent on water to reproduce, as are modern amphibians. The stem amphibians were not like today's frogs, toads, and salamanders—far from it. Many were monsters by comparison, quite lizard-like in shape, and most certainly they could not hop. They may, perhaps, have croaked.

Toward the end of the Carboniferous, the earth started to cool. The average air temperature was also about the same as it is today, which, relative to the long intervening Permian, Triassic, Jurassic, and Cretaceous periods, was comparatively cool. Over what was later to become Gondwana, the landmass that once joined Africa, Antarctica, Australia, Madagascar, India, and South America together, there existed a massive glacier. This part of Pangaea lay over the South Pole. The climate of the Late Carboniferous was marked by the waxing and waning of the size of the southern glacier during what is known as the Carboniferous Ice Ages.

As the southern glacier locked up water during the colder phases, the sea levels dropped, which created extensive lowland swamps. During the warmer interglacial periods the sea levels rose and shallow inland seas covered landmasses. The trees that grew at the time in the swamps were different from modern trees. They bore very high levels of lignin in their thick bark. Lignin is a nasty, toxic compound that protects the tree from being eaten by herbivores, insects at that time. Lignin is virtually indestructible.

The bacteria and fungi that cause modern bark to decompose had yet to evolve, so, once the ancient trees had fallen, their bark persisted in the oxygen-poor swamp beds as peat. It stayed there for millions of years and ultimately, under compression, formed coal.

Relative to modern levels, the carbon dioxide trapped as carbon in the coal depleted the atmospheric carbon dioxide and pushed up the oxygen levels. The sheer number of living trees using carbon dioxide and releasing oxygen added to the extreme ratios of the two gases.

Exactly how much oxygen was added to the atmosphere, and when, has been highly debated over the decades. Yesterday's perception was that the percentage of oxygen in the air during the Carboniferous was more than 10 percent higher than it is today. However, Sandra Schachat from Stanford University and her colleagues have updated the models that employ isotope analyses to estimate the oxygen percentages and shown that, whereas the atmosphere of the Mississippian period (358.9–323.2 million years ago) of the Carboniferous was indeed hyperoxic (with high oxygen concentrations, about 30 percent), in the last part of the Carboniferous, the Pennsylvanian period (323.2–298.9 million years ago), the oxygen concentration was virtually the same as it is in our modern atmosphere (about 21 percent).[1]

It was also yesterday's consensus that a hyperoxic atmosphere allowed some insects and stem amphibians to become giants, in what has been called Paleozoic gigantism. The extinct Meganeuridae are an example that is most frequently mentioned. They were a family of insects that looked very much like, and are closely related to, modern dragonflies. But they were giants, relatively speaking. With a wingspan of 71 cm (28 in), they dominated the air because there were no other aerial predators to hunt them. Jon Harrison from Arizona State University and his colleagues suggested that hyperoxia might have allowed insects to achieve larger sizes with smaller proportional tracheal volumes.[2] Matthew Clapham and Jered Karr from the University of California at Santa Cruz measured the dimensions of 10,500 fossil insect wings and matched the patterns to atmospheric oxygen levels from the Paleozoic to the end of the Cretaceous.[3] They showed that there was a good match for the first 150 million years of insect evolution, until the emergence other flying animals—birds in the Jurassic and bats in the Cenozoic—after which the match falls apart. These authors suggested that, before birds and bats started to compete with insects for aerial resources, insect gigantism was indeed molded by hyperoxia, especially during the Late Paleozoic.

Some stem amphibians also grew to massive sizes relative to today's frogs, toads, and salamanders. Some reached six meters in length. Some were aquatic, others semi-aquatic, and some were even terrestrial, unexpected for animals that evolved from fish. Many were crocodile-like in shape and function. In addition to breathing with their lungs, they exchanged respiratory gases across their skins, provided that it remained moist.

But with regard to insects, the Schachat study has also questioned Paleozoic gigantism because the authors argue that insect flight evolved only *after* the hyperoxic spike of the Mississippian period. In other words, flying insects evolved during the Pennsylvanian period, when the oxygen concentrations were about 21 percent. The study mentions Clapham and Karr's alternative explanation that habitat fragmentation, which I discuss further below, may have led to the evolution of novel forms and functions in insects that commenced to exploit emergent and vacant terrestrial niches.

Oxygen is a toxic gas, and under hyperoxic conditions it causes oxygen toxicity, which can be fatal. So it was always a problem for physiologists, including myself, trying to understand how Mississippian period animals coped with hyperoxic conditions. From the Schachat study we now know that flying insects simply did not exist at that time, as the so-called tetrapod gap in the fossil record reflects.[1]

Despite their ability to conquer land, Carboniferous stem amphibians were still, in some way or another, dependent on water. They needed to lay their eggs in water because the jelly-covered eggs were not nearly as waterproof as, for example, chicken eggs. They needed to remain within reach of water to breed. But again, as I discuss later, there is some contention about this topic, because some stem amphibians may have laid more derived (more highly evolved) eggs.

Stem amphibians were highly vulnerable to desiccation because they had moist skins that aided respiration. This constraint dramatically limited their potential to colonize the new terrestrial habitats that were about to emerge. The risks of high rates of water loss probably also explain their apparently larger body sizes. As I explain later, a large body size has the massive benefit of having a low surface area relative to its volume compared with smaller bodies, which reduces water loss from the skin.

Then the climate got extreme. The environment switched from warm and humid to cool and dry. The southern glacier that locked up the earth's water and lowered sea levels drove the change. The great drying and cooling at the end of the Carboniferous, about 305 million years ago, is called the Carboniferous Rainforest Collapse. The coal forests fragmented. They became surrounded by seasonally dry scrublands that imparted to the Late Carboniferous landscape a mosaic of new habitats—and evolutionary opportunity. And indeed, with the demise of the rainforests and swamps came a dramatic reduction in amphibian diversity, and with it evolutionary opportunity for new vertebrate forms. This was the first climate change event to start molding early tetrapods into today's two great lineages of furry and feathered endotherms, but it certainly was not the last.

An analogy of habitat fragmentation is described eloquently in the opening sentences of David Quammen's *The Song of the Dodo*, which is essential reading.[4] I cannot emulate Quammen's heroic, witty, journalistic writing style, but I do share his passion for popularizing science. Titled "Thirty-Six Persian Throw Rugs," Quammen's opening chapter invites the reader to imagine what happens when a three-by-five-meter (9.8 × 16.4 ft) Persian carpet is cut up into thirty-six equal rectangles. The result is not thirty-six cute little Persian rugs but "three dozen ragged fragments, each one worthless and commencing to come apart."

An ecosystem, he emphasizes, is a "tapestry" of species and relationships, and unravels when sections of it are chopped away or become isolated.

The Song of the Dodo is a popularization of the theory of island biogeography, which, even for scholars of evolution, can be a daunting subject. The architects of the theory were two renowned American biologists, Robert MacArthur and Edward O. Wilson. The theory predicts that the initial fragmentation of an ecosystem has devastating effects on species diversity. Fragmentation can be natural, such as that following climatic cooling and drying, or humans can cause it, for example, by chopping up a rainforest into fragments. Shrunken levels of resources drive the extinctions that follow fragmentation. Ultimately, though, organisms rebound from these perturbations and adapt not only to the new shrunken spaces and food supplies but also to the new interfragment spaces. Rapid extinction is followed by the gradual emergence of new species. This is the trend that has occurred after all major extinction events. It is also a recurring theme in the evolution of the furry and feathered lineages. The biggest extinction events are unmistakably bound to pulses in the evolution of traits that have ultimately allowed for endothermy in birds and mammals.

The dramatic drying and cooling at the end of the Carboniferous sped up the evolutionary cost and benefit processes that led to the emergence of the ancestor of the amniotes (defined shortly). It was this amniotic ancestor that diverged into the sauropsids and the synapsids. There were many critical evolutionary innovations in the ancestral amniotes, but the evolution of a large egg that could be laid on dry land frequently took center stage. Animals that lay amniotic eggs do not need a larval stage that demands an aquatic environment for development. For the first time, the ancestral amniotes gained relative freedom from water. They could now start to colonize new dry lands in between the "ragged fragments."

Success on dry land depended on innovations and adaptations that minimized desiccation, water dependence, and other problems that came with living on land. Waterproofing of the whole body was critical. Cooling was critical. Getting rid of toxic nitrogenous wastes was critical. Getting rid of carbon dioxide was critical, perhaps pivotal. Locomotion was critical. Early amniotes could no longer rely on the neutral buoyancy of water to move around cheaply. Digesting new dry-land plants was critical.

Waterproofing, cooling, breathing, walking, running, food processing, and urinating were the new land-based requirements, and they came with costs in energy and fitness, not big costs initially, but new, added costs nevertheless. But the fitness benefits must have increased simultaneously to meet and exceed these costs; otherwise none of the new lineages would have appeared.

Robert Lynn Carroll from the Redpath Museum at McGill University has argued that those Devonian stem amphibians that could and did haul themselves out of the water onto mud banks were the first vertebrates to benefit from the radiant energy of the sun. He and two colleagues, Jason Irwin and David Green, mathematically modeled the heat gains and losses that would have occurred in a meter-long Late Devonian stem amphibian.[5] At an air temperature of about 30°C (86°F), compared with a water temperature of about 20°C (68°F), the model amphibian would have taken about five hours to reach a body temperature of 32°C (89.6°F) while basking in the sun. The thermal fitness benefits to the animal would have been gained through a ubiquitous physical process in the universe, the Arrhenius effect.

Svante August Arrhenius was a Swedish physicist who won the Nobel Prize in Chemistry in 1903 and again in 1905. He is best known for his equation that predicted the dependence of temperature on chemical reaction rates. When heat is added to a living body it increases the kinetic energy of molecular motion. The process is unavoidable. Sometimes it is detrimental, even deadly, but sometimes it is beneficial. Heat increases the energetic costs of existence. It is deadly when the animal cannot find sufficient food to meet these heat-induced increases in the metabolic rate. The increase in metabolic rates means an increase in the rate at which carbohydrates, fats, and proteins are metabolized. It also increases the rate at which the energy product of these foods, adenosine triphosphate (ATP), is used to fuel metabolic processes.

A simplified generalization of the Arrhenius equation is called Q_{10}. It is a measure of the rate of increase in a physiological process with a 10°C (18°F) increase in temperature. In modern animals, the Q_{10} value of the metabolic rate falls somewhere between two and three. So, with an increase in the body temperature of the animal of 10°C (18°F), the metabolic rate would increase by somewhere between two and three times. The Arrhenius effect was a driving influence during the evolution of the furry and feathered lineages, but the first tetrapod benefactors were Devonian stem amphibians.

Stem amphibians started moving onto land at least 50 million years before the collapse of the coal forests. Let's call them the Devonian haul-out stem amphibians to differentiate them from later Carboniferous stem amphibians that were infinitely more landwise. Some lineages were already "experimenting" with ways of lifting the body off the ground once they had crawled or wriggled out of the water. The earliest experiments were not very successful, because their bearers went extinct. Later experiments were more successful, and it was the innovations in these lineages that paved the way for the transition to a committed land-based existence in the earliest sauropsids and synapsids. This was the *über* step toward the evolution not only of the mammals and the birds, but also of sustained hot bodies.

Robert Carroll has proposed multiple fitness benefits to Devonian haul-out amphibians generated by Arrhenius effects from basking in the sun. The rates at which ingested food could be digested would have increased in proportion to the Q_{10} values, somewhere between two and three times. The rate at which reproductive tissue, such as eggs, would have developed would have increased at a similar rate. Warmer muscles would have facilitated sustained swimming and faster reaction times while catching fish when the animal returned to the water after basking.[6] Higher metabolic rates and rates of food consumption would have promoted faster growth and, most important of all in terms of fitness benefits, allowed the earlier attainment of sexual maturity and enhanced rates of reproduction. More genes bearing the successful haul-out DNA codes would have been passed on to future generations. Natural selection would have been at work, powerfully, producing sun-powered benefits.

The oldest known fossil to illustrate a unique but ultimately flawed haul-out design is *Ichthyostega*, a fossil stem amphibian 350 million years old, from the Upper Devonian (Figure 2.1). The animal was reconstructed in 2005 by paleontologists from Uppsala University and Cambridge University, with some surprising observations in the structure of the vertebral column.[7] *Ichthyostega* is the first stem amphibian to show innovations in different parts of the vertebral column that enabled "walking" on land. These innovations involved the sizes, shapes, and orientation of the neural arches, which are the bony protrusions of each vertebra that extend upward from the spine. The authors of this study, published in the journal *Nature*, argued that it used its vertebral column as a suspension bridge to lift the mass

Figure 2.1. A reconstruction of the Devonian stem amphibian *Ichthyostega*, one of the first four-legged animals to haul out onto dry land. (Image by, courtesy of, and © 2012 Julia Molnar; the artwork was associated with Ahlberg, P. E., Clack, J. A., and Blom, H., "The Axial Skeleton of the Devonian Tetrapod *Ichthyostega*," *Nature* 437 [2005]: 137–140)

of the torso off the ground during a sort of shuffle-walk. The different orientations, lengths, and widths of the arches evolved to allow the attachment of the requisite muscles needed to arch the torso upward. No other stem amphibian at the time showed such stark differentiation of the vertebral column, and no other amphibian could move around on land, except for *Acanthostega*, another haul-out amphibian of the time. These were among the first tetrapods to benefit from the free energy provided by the sun.

Alas, the haul-out stem amphibians did not succeed. Their newly acquired benefits did not outweigh the costs.

About 50 million years after these early haul-out experiments, the Diadectomorpha appeared and lived close in time alongside the first amniotes—the synapsid "pelycosaurs"—at the end of the Carboniferous and in the Early Permian (shown in Figure A.1). These stem amphibians are called reptiliomorph amphibians because they looked like reptiles. They bore a close resemblance to the ancestral amniote in form and function. They were the sister group to the amniotes. Informatively, there were carnivorous, herbivorous, and omnivorous forms. I say informatively because the differences in features of the three feeding guilds within the same order allow for some of the earliest comparisons to be made about the respective innovations required to eat meat, plants, or anything. Importantly, also, they were all terrestrial. The diadectomorphs, collectively, were the major successful tetrapod conquerors of a rapidly changing world. Natural selection produced workable, successful haul-out designs compared with those of the Devonian haul-out stem amphibians.

Arguably the best "model" for comparisons with earlier stem amphibians and the first amniotes is *Diadectes*, the "crosswise biter" (Figure 2.2). It was the most derived of the diadectomorphs, the biggest, and the most herbivorous. *Diadectes* was about three meters long (9.9 ft) and *fully* terrestrial.

The solution to raising the body off the ground lay not in modifications of the vertebrae but, as is evident in *Diadectes*, in innovations in the hand, foot, arm, leg, hip, and chest bones. To conquer land successfully, tetrapods needed to be able to do push-ups. Push-ups are not possible when the arms, legs, hands, and feet point in the wrong direction. The earliest amphibians evolved from fish, and both had hands and feet adapted for swimming—they stuck out sideways.

Figure 2.2. A reconstruction of *Diadectes* by Dmitry Bogdanov. *Diadectes* was a fully terrestrial Early Permian reptile-like amphibian. (Image by, courtesy of, and © Dmitry Bogdanov, http://dibgd .deviantart.com)

Diadectes could do push-ups.

Diadectomorphs' hands and feet were on a level almost indistinguishable from those of early sauropsids and synapsids. The number of fingers and toes was reduced to five from the six or seven in the Devonian amphibians. Most mammals still have five fingers and toes. Some have fewer, but none have more. The number of carpal and tarsal hand and foot bones was also reduced, to three rows. These reductions increased the flexibility of the wrist and ankle and increased their ability to rotate. The last innovation that occurred was a change in the shape of the carpals and tarsals from rectangular to asymmetrical. The bones fitted each other like pieces in a jigsaw puzzle, which allowed for closer interlocking of bones in the hands and feet. These adaptations made push-ups possible; forward-pointing hands and feet were swung forward in an arc, in what is termed a sprawling gait. Importantly, though, the hands and feet landed closer to the center of the body, which profoundly increased the forward forces of the power stroke.

To be effective at lifting the torso off the ground, the limbs must be connected to the vertebral column. It is pointless to have arms and legs that can do push-ups and theoretically lift mass if they are not connected to the body that is being lifted. The paired fins of fish that became the arms and legs of stem amphibians and amniotes did not have any connection to the vertebral column—they floated around in the muscle below the vertebral column. The first connection between the limbs and the column was made in the Devonian haul-out stem amphibians, but it became vastly better in fully terrestrial reptiliomorphs like *Diadectes*. Bony connections were made between the hind limbs and the sacral region of the vertebral column, the pelvic girdle, and between the shoulder blade (scapular) and the rib cage, the pectoral girdle. These girdles paved the way for four-legged terrestrial locomotion.

Walking on land was only one of the adaptations required to conquer the terrestrial environment. Early amniotes also faced the massive problem of digesting plants. Prior to the coal forest collapse, stem amphibians were fish and insect eaters. They did not, and could not, eat plants. The capacity to eat plant material requires a suite of adaptations that early stem amphibians did not have: specialized teeth, jaws, and guts.

Diadectes ate plants, but before I discuss the evidence, I need to introduce the physiological challenges of eating plants.

Plant food falls into two categories: the contents of the cells and the structural walls of the cells. The cell contents are easy to digest because they contain the usual carbohydrates, simple ones like sugars, plus fats and proteins, mixed with some juice. But the cell walls are not easy to digest because they are made of cellulose and no juice. Cellulose is a carbohydrate, a complex one. It is a chain of glucose molecules strung together with very tough chemical bonds—glycosidic bonds. Only one enzyme, cellulase, can break these bonds. Bacteria and protozoans make their own cellulase, and so do fish moths (silverfish). That is why fish moths can eat your books. Very few mammals and birds possess cellulase, yet many of today's mammals rely almost exclusively on cellulose as their primary food source. Cellulose is the most abundant food source on earth.

Microbes break down cellulose, not into its constituent glucose units, but into fatty acids and methane, in a process of fermentation. Fermentation takes place in the guts, and also in the stomachs of ruminants. The methane is belched or farted out to add to the other greenhouse gases in the atmosphere. Modern ruminants—antelopes, cows, buffalos, and so on—have the most highly evolved stomach for fermentation, a four-chambered stomach. Chamber one, the rumen, ferments the plant material. Chamber two, the reticulum, sorts the fine stuff from the course stuff. The course stuff is regurgitated back into the mouth and rechewed—rumination—which is very effective at breaking down resilient material into smaller pieces with higher surface-area-to-volume ratios (discussed later) that can be attacked by microbes. Chamber three, the omasum, extracts the water so that it can be recycled via the salivary glands as drool mixed with bicarbonates. Chamber four, the abomasum is, well, exactly like our stomachs. It is pink, acidic, and digests everything, but especially proteins.

The four-chambered modern stomach of ruminants is the Rolls-Royce of fermentation structures in herbivores. It is the culmination of about

300 million years of evolution. Animals that switched from eating fish and insects at the end of the Carboniferous did not possess such complex stomachs. Rudimentary fermentation chambers were new innovations, yet they were a fundamental requirement for the shift to herbivory. The switch from eating fish and insects to carnivory—that is, eating anything meaty—was easy for early amniotes because it did not require complex alterations of the digestive tract. Meat is primarily protein, and as long as the stomach is acidic, protein digestion with enzymes can commence. All Carboniferous amphibians had acidic stomachs for digesting fish and insect proteins.

Proteins are broken down into their amino acid building blocks. When the amino acids are digested, the nitrogen in the molecule in its simplest non-gaseous form is ammonia, and it must be excreted. Ammonia is toxic if allowed to build up to high levels in the blood. Animals such as fish and larval amphibians (tadpoles) can easily get rid of ammonia molecules via the gills. Ammonia is a highly soluble small molecule that easily crosses the blood-water barrier. But this cannot occur in land-based animals. They need to convert the ammonia into a molecule that is not toxic and can be excreted not via the respiratory system but via the kidneys.

Uric acid is a nitrogen-based molecule produced to be excreted, which was essential for the conquering of dry land. It is not very soluble, especially in an acidic medium such as urine, and easily precipitates out as a white paste. Unfortunately, its capacity to precipitate out under high concentrations in the blood as well is the cause of the most painful affliction that humans suffer from—gout. Uric acid has a needle-like crystalline structure, and its favorite site for precipitation is where circulation is the slowest, in the left big toe. I have had personal experience with this.

The conversion of ammonia and other nitrogen products, such as purines, into uric acid is energetically costly, but the cost was clearly outweighed by the enormous benefits that being able to excrete uric acid conferred on the first conquerors of land.

Uric acid solved two massive problems in early egg-laying land dwellers. First, it solved the problem of how to get rid of toxic ammonia in the confined, sealed space within the shell of the amniotic egg. During development of the embryo in the egg, uric acid is stored as an insoluble white paste. If you examine the inside of the eggshell of a newly hatched chick, you will see this deposit. Second, uric acid does not need to be excreted with water in the urine as urea does. This chemical capacity means that animals that produce uric acid as their primary excretory product are much less dependent on water. Today animals in the sauropsid lineage, birds and reptiles, produce uric acid. It is the white paste that can be seen in their feces. Dinosaurs would have had white-streaked feces, too. In the cloaca, the uric acid precipitates out and is mixed with the feces from the digestive system. The overall evolutionary benefits of uric acid were that they allowed amniotes to colonize the dry scrublands that occurred during the Late Carboniferous and the Early Permian.

The furry lineage generates urea as the preferred nitrogen waste product. Urea is not toxic provided that it is well diluted. But again, there are two bitter costs associated with the urea excretory pathway: the energetic biochemical cost of converting ammonia to urea, which is greater than that of uric acid production, but, more important, the fitness cost in terms of the amount of water that is needed to dilute the urea and is lost from the animal in the urine.

Apart from the chemistry of herbivory, land-based vertebrates also required anatomical structures that would allow them to eat plants successfully. The evidence for herbivory in stem amphibians can be found in both the skull and the postcranial skeleton (the skeleton excluding the skull).

The postcranial features are a large body size, about a meter long, and expanded body cavities to house the expanded digestive tract needed for the fermentation of plants.

Diadectes had these.

In the skull the modifications associated with increasing herbivory were the teeth, lower jaw, and muscle insertion areas on the skull. The stem amphibians had rows of homodont teeth on both their upper and lower jaws. Homodont means that they were all the same in size and shape. A heterodont animal has teeth that are not all the same. We are heterodont animals. We have incisors, canines, premolars, and molars, each of which has a different role. *Diadectes* was one of the first true heterodonts. It had spatula-like front teeth perfect for nipping off leaves, and it had broad hind teeth perfect for grinding up leaves. These teeth were the precursors of mammalian incisors and molars.

Of course soft tissues do not fossilize, so we do not know where in the digestive tract of *Diadectes* fermentation would have taken place. It is highly likely that it did *not* occur in the stomach, because that would have compromised the animal's ability to digest proteins in an acidic stomach. The microbes essential for fermentation do not like acidic environments. The Devonian haul-out stem amphibians did not make the jump to herbivory instantly. Adding plant material to a diet of fish and insects took place gradually. Even *Diadectes*, the most herbivorous of diadectomorphs, would not have ignored high-quality food such as insects and fish, just as you'd grudgingly pass up a slice of blueberry pie to eat brussels sprouts. There was a continuum in the capacity to eat plants in addition to fish and insects. But, wherever diadectomorphs ended up on this plant-commitment continuum, all would have retained an acidic stomach to process quality food. The fermentation had to have been somewhere downstream from the stomach, probably somewhere in the colon, where many fermentation chambers occur in modern endotherms.

Tetrapods can do up to six different things with their teeth: hold prey, pierce prey, slice prey, crush bones, nip off choice bits, and grind up plants. The best features of tetrapod fossils, which identify them as insectivorous, piscivorous, nectarivorous, carnivorous, omnivorous, or herbivorous, are the arrangements and shape of their teeth. If there is evidence of tooth wear caused by tooth-on-tooth movement, it is often telltale evidence of the oral processing of plants. Effective chewing can occur only if the upper and lower teeth come into contact with each other, in what is called tooth occlusion. Tooth occlusion is not necessary to eat fish and insects, but it is certainly beneficial for processing plants. *Diadectes* and its kin were the first tooth occluders, and this crucial dental characteristic was passed on to the earliest amniotes via a common ancestor.

During the long course of mammal evolution, the bones that make up the lower jaw have shown a gradual change from four elements in reptili-

omorph amphibians—articular, angular, surangular, and dentary bones—
to one (dentary) in modern mammals. The dentary is the tooth-bearing
bone. The angular and the articular were not lost—they moved into the
ear to become the hammer (malleus) and the tympanic annulus, the bone
that houses and encircles the eardrum. The quadrate bone in the skull was
once the site of the hinge with the lower jaw, and it, too, moved into the ear
to become the anvil (incus). The hammer and the anvil joined the ancient
tetrapod hearing bone, the stapes, to form the three ossicles of the modern
mammalian middle ear. In the synapsid lineage, the surangular bone fused
with the dentary bone, but in the sauropsid lineage it had more specialized
functions. Indeed, true mammals are classified today as those synapsids that
have one bone in the lower jaw only, the dentary. So changes in the lower
jaw were associated not only with eating plants but also with more sensitive
hearing.

Increases in the size of the dentary bone relative to the articular and
angular bones along the synapsid lineage indicated a lower jaw that could
generate more power and greater biting and crushing forces. A strong jaw
was essential for eating plants. Early stem amphibians snapped at their food
and gulped it down, just as modern frogs do. Plant eaters cannot rely upon
such rudimentary ways of feeding.

Diadectes could chew.

Chewing was a huge evolutionary leap in food processing capacity. Remember, if these stem amniotes were to benefit from the Arrhenius effects
generated by free heat from the sun that elevated their metabolism, they
needed to be able to find and *process* sufficient food to meet the elevated
energetic demands of being hot. They needed to fuel the benefits of free
energy from the sun. Processing food was critical. For the first time, tetrapods could move their lower jaw not only up and down but also forward and
backward. This capacity involved a modification of the jaw joint.

Another feeding feature that appeared in *Diadectes*, often considered one of
the crucial starting requirements for being endothermic, was the secondary
palate. If you run the tip of your tongue along the roof of your mouth from
front to back, you are tickling your secondary palate. It is an enormously

useful feature of the skull because it separates the nasal cavity from the mouth or the buccal cavity. Early stem amphibians did not have a secondary palate. They had no choice about breathing through either their nostrils or their mouth. The nostrils opened into the buccal cavity, so breathing was via the mouth or nothing. *Diadectes* did have the choice, and, for the first time in tetrapods, animals could eat *and* breathe at the same time. They had not a full secondary palate, but a partial one. It was, nevertheless, the start of the separation of the airway between nose and mouth.

The secondary palate also increased food-processing efficiency by strengthening the snout, which is the primary explanation for its independent evolution again later in the furry lineage and in some members of the scaly lineage. It is most certainly not atypical that similar features should appear in separate lineages independently in response to similar selection pressures. Several sauropsids that showed crocodile-like snouts exhibited secondary palates, such as the derived crocodyliforms themselves;[8,9] anapsids within the Bolosauridae, such as the earliest bipedal tetrapod, *Eudibamus cursoris*;[10] the archosauriform phytosaurid *Machaeroprosopus lottorum*;[11] and even the theropod, spinosaurid dinosaurs.[12,13] In these forms with long, narrow snouts, the secondary palate resisted the bending and torsional forces associated with feeding. The secondary palate also allowed aquatic crocodile-like animals to be sit-and-wait predators because they could breath with only their nostrils above water.

Early stem amphibians had small lungs and a rather inefficient way of ventilating them. They used buccal pumping, the up-and-down movement of the floor of the mouth cavity. The moist skin of stem amphibians was a pivotal supplement to respiration, especially for getting rid of carbon dioxide. But this method of dual respiration was useless for living on dry land. A skin that was not waterproof would have led to death from water loss through evaporation within hours if the lost water could not be replaced. The skin needed to be waterproof, which meant its respiratory capacity had to be relinquished to a degree. This prerequisite placed enormous demands on small lungs ventilated by buccal pumping alone. A new pumping system and expanded lungs were needed to compensate for the loss of skin respiration but also to accommodate the new increased metabolic rates driven by Arrhenius effects during the heat of the day.

Again, a staggering number of new anatomical adaptations appeared in a relatively short period of time. The new pump exploited the rib cage and its associated muscles. Expanding and contracting the rib cage could draw air in and out of the lungs. This is called costal (rib) pumping. We use costal pumping. The external and internal intercostal muscles between the ribs cause expansion and contraction of the thoracic cavity and hence inhalation and expiration, respectively. The ubiquitous tetrapod "core" muscle, the *transverse abdominis*, which sulks under your six-pack muscle, the *rectus abdominis*, assists the process. To work efficiently, costal pumping required quite a number of modifications—in head shape and size, neck length, and rib cage.

Christine Janis from Brown University and her colleague Julia Keller have argued that the evolutionary pressure on the respiratory system was so strong during the shift to a terrestrial existence by the amniotes that its importance was equal to, if not greater than, that of the evolution of the amniotic egg.[14] The ideas that these authors present hinge on one of biology's most important observations: the relationship between an object's volume and its surface area (SA:V) as it gets bigger. It does not matter whether we are dealing with atoms, molecules, organs, whole bodies, territories, ecosystems, or universes—the concept remains the same: as geometrical size increases, the volume and surface area of an object do *not* increase in the same ratio. A doubling of length, width, or height is not accompanied by a doubling of surface area or volume. This is because two- or three-dimensional measures, such as surface area and volume, do not increase in the same proportion as one-dimensional measures, such as length. Large objects with shapes geometrically similar to those of smaller objects have a smaller surface area relative to their volume than do the smaller objects (Figure 2.3). For example, a cell with a larger diameter has a smaller surface area relative to its volume than does a smaller cell.

This SA:V concept has a colossal influence on the flux of matter to and from cells, organs, and whole animals. The most important fluxes involve gases, such as carbon dioxide and oxygen, but also water and heat. In later chapters I will discuss how these fluxes have a determinate influence on the

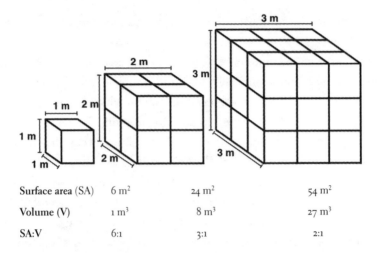

Surface area (SA)	6 m²	24 m²	54 m²
Volume (V)	1 m³	8 m³	27 m³
SA:V	6:1	3:1	2:1

Figure 2.3. The comparative surface area of three geometrically similar-shaped squares of different sizes. The surface-area-to-volume ratio of the smallest square is three times that of the largest square. (Graphic by the author)

metabolism required to sustain a warm body. Small animals lose and gain heat far faster than larger animals. They also exchange gases more readily than large animals. From a terrestrial perspective, though, whereas gases and heat are two-way fluxes, water is almost always only lost from an animal, and this occurs faster per unit of body mass in small animals than it does in large ones.

Janis and Keller argue that acquiring oxygen is not the problem. The real challenge is in getting rid of carbon dioxide. Oxygen is exchanged in the lungs much more readily than is carbon dioxide. Most carbon dioxide is exchanged across the skins of stem amphibians. If carbon dioxide cannot be offloaded fast enough from the blood, it causes the blood to become acidic, creating acidosis. It changes the acid-base balance of the blood because of the buildup of carbonic acid formed in the red blood cells from carbon dioxide and water. Carbonic acid undergoes a reversible reaction to form bicarbonate ions and hydrogen. The bicarbonate ions transport the carbon dioxide to the lungs or the skin's surface, where they are again converted to carbonic acid and back to carbon dioxide and water. Acidosis is lethal but can be avoided to some extent by increasing the bicarbonate ions in the blood as a sort of buffer, an adaptation to respiratory acidosis. Most amniotes possess this adaptation that evolved in the shift to land dwelling.

Janis and Keller propose that the problems associated with getting rid of carbon dioxide remained the biggest hurdle to shifting to land. They argue that the evolution of costal pumping was the solution.

Buccal pumping can be optimized with a large, broad buccal cavity. Modern frogs retain this trait to this day. Buccal pumpers do not have a trachea connecting the mouth to the lungs, so the neck cannot be long. A neck has never evolved in frogs. Buccal pumpers do not need ribs to pump air. All frogs lack ribs. These features offered Janis and Keller evidence of rapid changes in the head and its link to the rest of the skeleton that were associated with the shift from an aquatic to a more land-based existence.

Diadectes is again a useful ancestral amniote "model" to illustrate these modifications. It had a longer neck, smaller head, and a bigger rib cage than, for example, *Eryops*, the earlier, less terrestrial, haul-out stem amphibian. It also had narrower ribs to accommodate the intercostal muscles needed for pumping air in and out of the lungs.

The problem of desiccation was paramount for early tetrapod land dwellers. They could not retain the wet, slimy skin of their forebears, which was so effective in exchanging respiratory gases. The skin needed to be more waterproof—much more waterproof. The outermost layer of stem amphibian skin, the epidermis, is thin and soft and filled with mucous glands that keep the skin moist.

The outer layer of the skin epidermis is called the stratum corneum. A new form of fibrous keratin, beta-keratin, together with lipids, evolved in the stratum corneum that, respectively, made the skin stronger and made it considerably more waterproof. Lipids are the fundamental barriers to desiccation and water loss from the skin in land animals.[15] The more lipids there are in the stratum corneum, the more waterproof the skin. Most modern amphibians do not have a waterproof skin because their stratum corneum contains alpha-keratin only, which is much softer than beta-keratin. Later, in the evolution of the modern mammals and birds, beta-keratin became the hard structural protein in claws, nails, and feathers.

The evolution of the amniotic egg from the eggs of stem amphibians—anamniotic eggs—was an extraordinary evolutionary innovation illustrating how powerful natural selection was at that time in producing new land-based animals. All tetrapods that inherited the amniotic egg are known as amniotes—the furry, scaly, and feathered lineages.

A developing embryo within an egg needs to accomplish three things: obtain nutrients, exchange gases, and excrete waste products. The stem amphibian egg was very simple compared with the amniotic egg. It was an embryo attached to a yolk sac that supplied the embryo with nutrients. Surrounding the yolk was a vitelline membrane that was, in turn, surrounded by a mass of jelly that was mostly water, absorbing it when in water but highly desiccant when on dry land. The respiratory gases were exchanged with the surrounding water through the jelly, but the jelly had a poor capacity for gas diffusion that placed a limit on the size that the neonate, and hence the adult, could attain. The ancestral anamniotic egg therefore severely limited the capacity for stem amphibians to attain larger sizes, and this remains the case in modern amphibians. And yet stem amphibians such as *Diadectes* did get big, so how was the gas diffusion problem overcome?

Mary Packard from Colorado State University at Fort Collins and Roger Seymour from the University of Adelaide proposed a sequence of evolutionary events leading from the jelly-covered anamniotic egg to the amniotic egg.[16] "The key innovations that permitted the evolution of the larger-reptilian eggs," they suggest, "were the envelopment of the egg by a fibrous shell membrane external to the jelly layers and transfer of water from the jelly layers into the yolk compartment." The shell provided structural support to the egg, and the transfer of water from the jelly layers to the yolk eliminated the problem of gas diffusion and dried out the spaces between the fibers of the shell membrane. Both of these innovations permitted rapid diffusion of gases between the embryo and the environment, which allowed eggs at this stage to become bigger. Bigger eggs have a lower SA:V that provides a greater relative volume in which to store the moisture necessary for complete embryo development.

Then three new membranes appeared, the allantois, chorion, and amnion (Figure 2.4). They did not appear out of nowhere. They developed from the three germ layers in the embryo, the ectoderm, mesoderm, and endoderm.

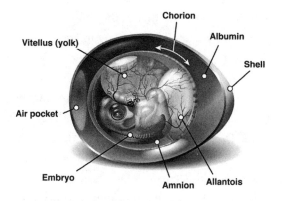

Figure 2.4. The amniotic egg. (Image modified from KDS4444, CC 4.0, https://en.wikipedia.org/wiki/File:Chicken_egg_diagram.svg)

The allantois membrane forms a sort of balloon in which the narrow inflating bit is attached to the digestive tract of the embryo and the bulgy bit squashes against the inner surface of the chorion, which is the membrane lining the inner surface of the shell. The joined parts of the highly vascularized allantois and the chorion act as a sort of primitive placenta, an area of the shell where the embryo communicates in a controlled manner with the outside world in exchanging gases, oxygen and carbon dioxide. The balloon space of the allantois is ultimately filled with waste products, such as uric acid. The amnion membrane surrounds and protects the embryo and provides an aqueous medium, the amniotic fluid, in which the embryo can develop. The amniotes (Amniota), which include the modern birds and mammals, take their name from the amnion membrane. Amphibians and fishes, the Anamniota, do not have an amnion. The space between the amnion and the chorion is filled with albumin, the egg white.

The three egg membranes that evolved in the amniotic egg were passed on to the furry lineage, even the placental mammals that do not lay eggs. In placental mammals, the fused chorion and allantois are thrown into tight folds that form the placenta. The placenta in the uterus is the organ that allows the exchange of gases and nutrients between the developing embryo and the mother. Indeed, in viviparous reptiles that bear live young, the joined allantois and chorion act as a placenta allowing for gas exchange between the mother's system and that of the embryo. In viviparous lizards and snakes, the eggs lack a shell and are retained in the uterus until the embryo has fully developed.

The evolution of the amniotic egg must have occurred before the emergence of the amniotes as we define them, because how else would stem amphibians such as *Diadectes* have attained such a large body size? Gas exchange facilitated by the chorion and allantois allowed eggs to attain ever larger sizes, and embryos could develop fully without having to undergo an aquatic larval stage, such as occurs in modern frogs.

In the final development in the amniotic egg, calcium was laid over the eggshell membrane in the oviduct before the eggs were laid. Calcium added considerable strength to the eggshell and provided the calcium necessary for bone development in fully developed neonates. A hard-shelled egg was also more waterproof and could be laid on land. The amniotic egg at this point became the cleidoic egg. So all amniotes that lay hard-shelled eggs, such as birds, lay cleidoic eggs.

The dramatic climate changes at the end of the Carboniferous spelled the dawn of the tetrapod land dwellers. The synapsid lineage that ultimately led to the mammals split into two major groups in the Late Carboniferous, the "Pelycosauria" (*Dimetrodon* and *Sphenacodon*) and the Therapsida (shown in Figure A.1). The Pelycosauria have always been a problematic group for taxonomists because they are paraphyletic. This means that the known species of extinct "pelycosaurs" cannot be arranged into a single tree radiating from a single ancestor. Pelycosauria is a sort of taxonomic trash can of stem synapsids that cannot be sorted into acceptable family trees. The name pelycosaur is inappropriate for the five families of nontherapsid synapsids anyway because the name is derived from features that occur in some species of two families only. The most appropriate term for these five families should be "stem synapsids," which is what I will call them.

Although many boxes had been ticked to make conquering land successful, the stem synapsids completed the checklist. They dominated the earth during the Permian. They were adapted to move away from water, although it is unlikely that they were ever totally independent of it for long periods of time. Given their large body size, they must have laid amniotic or cleidoic eggs away from water, and they are highly likely to have defended their nests as crocodiles do today. They also must have exhibited a waterproof skin dominated by keratin-lipid complexes. From their fossils it is known that they had well-developed pectoral and pelvic girdles. They could do push-ups. They could walk on land, with a waddle, by swinging their legs out sideways in what is called a sprawling gait. Some species, such as *Dimetrodon*, possessed grotesque extensions of the vertebra to form "sails," the function of which remains highly speculative (Figure 2.5). More highly derived stem synapsids, such as *Heleosaurus*, displayed postures and foraging behavior very similar to those of modern monitor lizards.

Cotylorhynchus was a notable stem synapsid because it had very large nostrils and a snout. It was about 6.5 m (21.3 ft) long and had a tiny head relative to the size of its bulky body. In modern mammals large nostrils are

Figure 2.5. A reconstruction of the iconic sail-back stem synapsid *Dimetrodon gigas*, a dominant carnivore of the Permian, together with the haul-out stem amphibian *Eryops megacephalus*. (Image by, courtesy of, and © Dmitry Bogdanov, http://dibgd.deviantart.com)

normally associated with high rates of airflow, such as occurs in pronghorns at full gallop, or the retention of respiratory water, through nasal cooling (Appendix 2). *Cotylorhynchus* could not gallop. It was too bulky and had a sprawling gait. However, the large nostrils may have minimized water loss from the animal during the daytime, which would have allowed it to venture farther away from water.

The stem synapsids are generally considered to have been ectotherms, but this need not necessarily have been the case. As more and more is learned about the evolution of endothermy and as it becomes easier to join the dots with the emergence of new data, the older the origin of endothermy becomes. We should therefore not discount the prospect of endothermic features having appeared for the first time in the stem synapsids. Nevertheless, innovations evolved in them upon which natural selection could act to ultimately produce endothermic tetrapods. It was from a stem synapsid of the Sphenacodontidae that the therapsids, the next major group of archaic mammals, evolved, and they did indeed exhibit the first traits that were truly indications of endothermy. They are colloquially known as "mammal-like reptiles" and belong to a well-recognized monophyletic group of stem mammals called the Therapsida, named by Robert Broom. But they are not reptiles, so I call them just therapsids. (According to one argument, the term "reptile" refers to all sauropsids.)[17] And the best place to find the highest diversity of therapsid fossils is in the Karoo Basin in South Africa.

THREE

The Karoo

I would like to tell you about the Karoo, because in its shales and sand-stones lie fossils that reveal many secrets that we need to know about if we wish to understand why mammals and birds are endotherms. The Karoo holds the best fossil record in the world of the early ancestry of mammals, in particular, and also what happened to them, so very dramatically, at the end of the Permian, about 252 million years ago. Its fossils span a continuous record in ancient time from the Middle Permian to the Middle Triassic. The Karoo's record of the epic extinctions and originations over this time have allowed us to piece together parts of the puzzle of the evolution of endothermy, certainly in the therapsids.

The Karoo is the remnant of a vast inland 600,000-km^2 (372,822-mi^2) basin in South Africa bounded in the south and west by the formidable Cape Fold mountains. It was formed at the same time as the split between the synapsid and sauropsid lineages, around 320 million years ago. At that time the part of the supercontinent Pangaea that would eventually become the Karoo lay over the South Pole. So it was, initially, a very cold place. Glaciation during this cold phase deposited on the basin floor unsorted debris (tillite) a kilometer thick, which became the lowermost formation of the Karoo Supergroup—known as the Dwyka Group. As Gondwana drifted northward, the Karoo became warmer and wetter and turned into a vast inland sea with extensive swampy deltas on its northern shores. It was at this time that the coal deposits on which South Africa relies heavily as its primary source of energy were laid down. These sedimentary layers, which contain a rich fossil assemblage of *Glossopteris* plants, now make up the Ecca Group. Overlying the Ecca rocks are the Beaufort and Stormberg Groups, deposited in a terrestrial environment, and they contain arguably the greatest deposits of therapsid fossils on earth.

Today the Karoo is no longer a sea—it is a desert. It is dry sheep country, home to the renowned Karoo lamb. This meat gets its delicate, spicy flavor from the diverse array of aromatic, woody shrubs that grow there. The Karoo is screwed up, ecologically. It was long overgrazed, abused not only by European farmers but also by indigenous Khoi nomads and their fat-tailed sheep before them. It is a beautiful landscape, but it is desolate and arid. It undergoes extremes of hot and cold, and the wind blows hard.

No story about the Karoo can be told without introducing the Bernard Price Institute (BPI) for Paleontology at the University of the Witwatersrand in South Africa, now replaced by the greatly expanded Evolutionary Studies Institute (ESI). The BPI has a long and distinguished history, and some of its early staff made watershed contributions to paleontology and paleoanthropology. I need to tell you about the legendary fossil hunters of the Karoo who were in some direct or indirect way associated with the BPI and who discovered and described many of the ancestors of the mammals that we will meet.

Named after its original benefactor, Bernard Price, the BPI was established in 1945 in response to the prompting and enthusiasm of the Scottish-born medical doctor-turned-paleontologist Dr. Robert Broom (Figure 3.1). Despite having been an anti-Darwinist throughout his career, Broom had undertaken a life quest to understand the origins of mammals. At the age of 31, Broom left Glasgow to study monotremes and marsupials in Australia for a few years and then went to South Africa in 1897 to hunt for fossils in the Karoo, where early mammalian progenitors had been discovered. Broom found many more and named them Therapsida.

Robert Broom was deeply troubled about the fate of the Karoo fossils. In 1937, in a paper in which he described *Procynosuchus delaharpeae*, the ancestral cynodont and hence at one time your ancestor, his opening paragraph is etched with anguish:

Figure 3.1. Robert Broom (left) with Peter van Riet Louw, renowned South African paleontologists. The photograph was taken in 1937 in the Cave of Hearths, Makapansgat Valley. Broom is holding a tooth of the extinct Giant Cape Zebra, *Equus capensis*, which he described and named. (Permission by, courtesy of, and © South African Archaeological Society, http://www.archaeologysa.co.za/gallery)

Few who have not visited South Africa can realize how extensive are the fossil deposits of the Karroo. . . . There are always a million fossils worth collecting lying exposed on the surface, and . . . of these a hundred thousand are weathered into dust every two or three years. In many places the shale weathers away at the rate of an inch a year, and every inch reveals hundreds of new specimens. Prof. Camp was in South Africa last year collecting for the California University. On one farm to which I directed him he collected over a hundred good skulls of *Dicynodon* in one day. In three months he made a collection that I think far surpasses the great collection in the British Museum.[1]

Broom was extremely eccentric. He carried out his fieldwork throughout his career dressed impeccably in a fine suit. It is rumored that when he got too hot, he stripped naked and continued to work. An all-or-nothing human typhoon, he studied the therapsids, named many new species, and ultimately wrote more than four hundred scientific publications based on the Karoo fossils.

Broom set up a medical practice in the Karoo town of Pearston in 1900 but spent most of his time searching for fossils. In 1903 he was appointed professor of geology and zoology at Stellenbosch University and a few years later the curator of fossil vertebrates at the South African Museum in Cape Town. Broom was completely reliant on a free railway pass he'd been granted to travel to the Karoo to undertake collecting trips whenever he could. Controversially, though, a government minister reversed his train pass, and he consequently resigned from Stellenbosch University in 1910. Over a period of seven years he had managed to publish a hundred papers on Karoo vertebrates. Broom's penchant for fossil hunting came at the expense of his medical practice, and he resorted to trading in Karoo fossils, paying finders peanuts and selling off the spoils to overseas collectors. He did not endear himself to the South African museum establishment.

Broom finally devoted himself to paleontology full time and by 1925 had published 250 papers and named about 70 new genera and 200 new species of Karoo vertebrates. However, during the depression of the early 1930s, Broom became destitute. He was rescued by an appeal from Raymond Dart to the South African prime minister, Jan Smuts, to create a position for Broom so that he could continue his work. Smuts duly arranged for Broom to be appointed keeper of vertebrate paleontology and anthropology at the Transvaal Museum in Pretoria, a position he held until his death on 6 April 1952 at the age of 84.

Many amateurs were inspired by Robert Broom to become fossil hunters. Notable amateurs were Sidney Rubidge, grandfather of the current director of the ESI (formerly BPI), Bruce Rubidge, and "Croonie" Kitching, an overseer of road construction in the Karoo. Sidney Rubidge farmed Wellwood near Nieu-Bethesda. Both Rubidge and Kitching passed on their enthusiasm for fossil hunting to their kids. Croonie had eight children—three sons became excellent fossil hunters—but it was one son, James William Kitching, who excelled, and whom Broom described as the "greatest fossil finder in the world."

The Kitchings also lived in the small Karoo town of Nieu-Bethesda, which became the hotspot for finding therapsids. At the tender age of 23, James Kitching was appointed as the BPI's first staff member and full-time technical officer. After returning from active service during World War II, he dived straight into the Karoo and within six months had collected several hundred well-preserved skulls, mostly of therapsids, around Graaff-Reinet and Nieu-Bethesda. Although James Kitching never held an undergraduate degree, the University of the Witwatersrand permitted him to register for an MSc and ultimately awarded him a PhD in 1972 for his thesis on the stratigraphic and geographic distribution of Karoo fossils.

Sidney Rubidge discovered the holotype of *Dinogorgon rubidgei*, a large carnivorous gorgonopsid therapsid from the Late Permian, and the fossil remains on Wellwood to this day (Figure 3.2). I went to see it in 2017 because the skull is stunning. Robert Broom named it and encouraged Sidney Rubidge to collect more fossils. This was the beginning of a fruitful

Figure 3.2. A reconstruction of *Dinogorgon rubidgei*, a carnivorous gorgonopsid therapsid from the Late Permian named after Sidney Rubidge by Robert Broom. (Image by, courtesy of, and © Dmitry Bogdanov, http://dibgd.deviantart.com)

collaboration that lasted until Broom's death. Indeed, Sidney Rubidge and the Kitchings continued to find and amass a large number of fossils and hoard them on Wellwood. This is the greatest private collection of therapsid fossils and is a testament to the toils and passion of amateur scientists. Today, in Nieu-Bethesda, you can visit Wellwood or the Kitching Fossil Explora- tion Centre and view these fossils, or you can wander through the veld and be guided to where the fossils lie, in situ, as Broom had described them, weathering away to dust.

Croonie was seriously wounded in action in Egypt during World War II, and, despite treatment in hospitals in Cairo, Durban, and Port Elizabeth, ultimately succumbed to his injuries on 4 January 1943 in Graaff-Reinet. Here is part of the eulogy given by Sidney Rubidge to a man who made an enormous contribution to understanding mammal origins:

> As a hunter and successful discoverer of fossil skulls he probably ranks first in S. African History. He started searching at the age of 16, and for the first 20 years sent his finds to Dr. Broom and to various Museums in S.A. and for the past 7 years collected for the Wellwood Museum. It would be difficult to arrive at the actual number of Typed S. African specimens, which should be credited to his own finding, at least 30 alone are in the Wellwood Collec- tion, but it is possible that a greater number than this are distrib- uted elsewhere and for which he did not receive the credit.[2]

James Kitching carried on the family obsession with distinction. Apart from spending extensive periods living in a tent in the Karoo in pursuit of fossils, he was also directed to hunt for fossils in the newly discovered limestone Pleistocene deposits at Makapansgat in the Mokopane (Potgie- tersrus) district. There he found his most notable Pleistocene fossil, the type specimen of the ape-man *Australopithecus prometheus*, described by Broom in 1947. This hominid was later considered the same as *Australo- pithecus africanus*, the famously controversial Taung Child fossil described in 1925 by Raymond Dart. British paleontologists Arthur Keith and Grafton Elliot Smith ridiculed Dart's work at the time because they simply would not accept that humans could have evolved anywhere except in the United

Kingdom. But following pressure by Broom and American paleontologists, Sir Arthur Keith finally vindicated Dart 22 years later, admitting that "Dart was right, and I was wrong." More recently the name *Australopithecus prometheus* has been reinstated.

Because of the extensive fossil record from the Permian to the Jurassic, the Karoo is the best place to study the evolutionary transition from therapsid to mammal. However, the earliest representatives of the synapsids were the stem synapsids, the so-called pelycosaurs, which are best known from Carboniferous and Early Permian deposits from Texas. Robert Broom was one of the early protagonists of the idea that stem synapsids gave rise to therapsids, an idea accepted today. However, the fossil record of the earliest therapsids is sparse.

Tetraceratops insignis, which means "four-horned face emblem," from Texas deposits, was originally considered the oldest therapsid, but a recently discovered fossil from China, *Raranimus dashankouensis,* described by Liu Jun and Bruce Rubidge, also shows both stem synapsid and therapsid features and is now considered the most ancestral therapsid.[3] In the 1980s Bruce Rubidge, then working at the National Museum in Bloemfontein, and his assistant John Nyaphuli, were responsible for finding the oldest therapsids from the Karoo, which by then already displayed a very diverse fauna. The Therapsida comprises six families, all of which are known from South Africa: Biarmosuchia, Dinocephalia, Anomodontia, Gorgonopsia, Therocephalia, and Cynodontia. The Cynodontia are the most derived therapsids and were the ancestors of mammaliaforms and true mammals.

The Dinocephalia were the first large vertebrate animals to live on land, 272–260 million years ago, and included both herbivorous and carnivorous species. *Tapinocephalus* ("humble head") was a massive herbivore and rather odd. It had a thick, bony swelling on its head, probably used for head-butting rivals, and it weighed the equivalent of a black rhino, about 2,000 kg (4,409 lb). Dinocephalians went extinct 260 million years ago,[4] and it was another 200 million years, in the Paleocene and Eocene, before mammals again become so big and so dominant.

The Dicynodontia, characterized by having a horny covering on their mouths, like a sort of beak, were the most successful therapsids of the Late Permian and Early Triassic (shown in Figure A.2). They were highly diversified and ranged in size from several tens to several thousands of grams. They were herbivorous. Apart from a pair of large tusks in some species, most dicynodonts had no other teeth; hence the name Dicynodontia. The Dicynodontia used their horny beaks to tear tough vegetation.

Dicynodont fossils have been collected in the Karoo since the mid-1800s. The famous English paleontologist Richard Owen named them from specimens sent to the British Museum of Natural History by collectors such as Andrew Geddes Bain before the Broom era. Owen named one of these dicynodonts *Oudenodon bainii* in honor of Bain. Bain was a prolific road-building engineer and built some of the most spectacular roads through mountain passes in South Africa.

Lystrosaurus ("shovel lizard"), named by the American paleontologist Edward Drinker Cope, is possibly the best-known dicynodont. The various *Lystrosaurus* species ranged in mass from 5 kg (11 lb) to 200 kg (441 lb); the biggest was about the size of a fat pig. The bigger species reached lengths of nearly 3 m (9.8 ft). Their front legs were more robust than their back legs, indicating that they were good diggers and lived in burrows. They were not at all cursorial (having limbs adapted for running fast)—they still walked with a sprawling gait.

Lystrosaurus is the most famous dicynodont for several reasons. First, specimens have been found in numerous countries in both the northern and the southern hemispheres that once made up the supercontinent Pangaea, countries such as South Africa, India, China, Mongolia, and European Russia, and also on the continent of Antarctica. Second, they were one of the few therapsids to survive the end-Permian extinction event and went on to completely dominate the post-extinction recovery phase of the Early Triassic; in so doing they earned the title "disaster taxon." Therapsids such as *Lystrosaurus* continued to dominate on earth throughout the Early Triassic until the emergence of more competitive reptiles from the sauropsid lineage during the Late Triassic (shown in Figure A.2).

James Kitching was the first person to identify *Lystrosaurus* fossils from Antarctica. He joined an expedition of Ohio State University's Institute of Polar Studies to the Queen Maud Mountains in 1970. The significance of finding *Lystrosaurus* fossils on Antarctica was that it confirmed the former continental link between southern Africa and Antarctica. Today a prominent rocky ridge (85°12′ S 177°06′ W) on the west side of Shackleton Glacier, between Bennett Platform and Matador Mountain in the Queen Maud Mountains, has been named and mapped as "Kitching Ridge" in his honor.

Apart from the South Urals Basin in Russia, the Karoo material contains the best evidence of the dramatic vertebrate changes that occurred between the Permian and Triassic geological periods—the time of the Permo-Triassic Mass Extinction, the greatest extinction of all time.

Jennifer Botha-Brink from the National Museum in Bloemfontein has studied the *Lystrosaurus* species turnover across the extinction boundary in the Karoo, where the *Lystrosaurus* fossils are most abundant.[5] She has shown that the larger, more specialized species, such as *Lystrosaurus maccaigi*, did not survive the extinction event. However, *Lystrosaurus curvatus* did; it was the great dicynodont survivor. It was the least specialized of the dicynodonts and became the most dominant vertebrate of the Early Triassic. But it did not persist far into the Triassic.

Two new species of *Lystrosaurus* emerged during the Early Triassic in the Karoo, *Lystrosaurus murrayi* (Figure 3.3) and *Lystrosaurus declivis*. They undoubtedly occupied different niches to those of *Lystrosaurus curvatus*. They were considerably smaller than *Lystrosaurus maccaigi*, which suggests that having a smaller body size might have been advantageous in the tricky environment of the Early Triassic. This is an important supposition, because decreasing body size was the driving force for enhanced endothermy during the Late Triassic (discussed in Chapter 6).

Burrowing was crucial to the success of the Early Triassic therapsids such as *Lystrosaurus*. It allowed them to avoid the heat of the day and thereby conserve water that would otherwise have been needed to keep the animals cool through evaporative cooling. Conserving body water was much more important for therapsids at this time than it was for archosauriforms,

Figure 3.3. *Lystrosaurus* species dominated the fauna of the Early Triassic. Illustrated here is *Lystrosaurus murrayi*, reconstructed by Dmitry Bogdanov. (Image courtesy of and © Dmitry Bogdanov, http://dibgd.deviantart.com)

the reptile clade more derived than but including Proterosuchidae. Therapsids needed to commit a certain amount of water to the excretion of urea, whereas archosauriforms did not because they excreted uric acid. So behavioral adaptations such as burrowing would have saved therapsids critical amounts of water and thereby have reduced their risk of being killed by predators while out searching of water.

The Karoo fossils also provide evidence of the first emergence of reptiles in the Early Triassic, which started to enter the story of the evolution of endothermy. These were archosauriforms, which were the ancestors of the archosaurs, which include crocodiles, dinosaurs, and pterosaurs (shown in Figure A.1).

The first archosauriform to dominate the Early Triassic landscapes was *Proterosuchus* (Figure 3.4), a crocodile-like creature that preyed on a variety of therapsids, especially dicynodonts such as *Lystrosaurus* species. It was about 2 m (6.6 ft) long and retained the ancestral sprawling posture. Robert Broom named the type specimen of *Proterosuchus fergusi* from a specimen found in the Karoo Basin.[6] *Proterosuchus* was initially thought to be aquatic or semi-aquatic because of its crocodile-like features and eyes high up on its head, which apparently would have exposed very little of its head while submerged. However, *Proterosuchus* does not share the bone density characteristics of typical aquatic vertebrates.[7]

The biggest carnivores of the Early and Middle Triassic were not therapsids, which, as I've intimated, were starting to undergo a miniaturization process; they were archosauriforms of the family Erythrosuchidae. Indeed, one species, *Garjainia madiba* (Figure 3.5), is named in honor of the great South African luminary Nelson Mandela.[8] Madiba is Mandela's Xhosa clan name. *Garjainia madiba,* as well as other closely related species such as *Erythrosuchus africanus,* described by Robert Broom in 1905,[9] and indeed also *Proterosuchus,* exhibited a telltale signature of an early endothermic status—juveniles grew fast.[7] Like the Early Triassic therapsids that I discuss in the next chapter, fast growth rates early in life in archosauriforms are thought to have evolved in response to the unpredictable climate of the

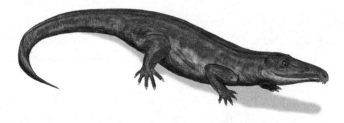

Figure 3.4. A reconstruction of the Early Triassic *Proterosuchus,* a fearsome aquatic and terrestrial carnivore. (Image by, courtesy of, and © Nobu Tamura, http://ntamura.deviantart.com)

Figure 3.5. A recently described Early Triassic carnivorous archosauriform, *Garjainia madiba,* named after Nelson Mandela. Its bones show indications of rapid growth, indicating an early spurt of endothermy in the sauropsids. (Reconstruction by, courtesy of, and © Dmitry Bogdanov, http://dibgd.deviantart.com)

Figure 3.6. Looking larger than it really was, a reconstruction of *Euparkeria africanus*, an agile derived archosauriform from the Middle Triassic that showed early evidence of cursoriality in the form of hind limbs longer than the fore limbs. (Image by, courtesy of, and © Cristóbal Aparicio Barragán, http://www .seirim.net)

Early Triassic.[10–13] *Erythrosuchus* was large. It was about 5 m (16 ft) long and 2.1 m (6.9 ft) tall. Its head alone was 1 m (3.3 ft) long, and it walked with a more upright stance than its ancestors. So the fauna of the Early Triassic was offering glimpses not only of the gigantism that was to come, such as that of the giant sauropod dinosaurs, which were the biggest land animals to ever occur, but also of early endothermic characteristics, namely, fast juvenile growth rates. But there was also miniaturization to come.

In 1913 Robert Broom named a curious archosauriform that appeared in the Karoo in the Middle Triassic about 240 million years ago *Euparkeria capensis* (Figure 3.6).[14] It was a small archosauriform, about half a meter (1.6 ft) long, but it displayed the first deviation from a typical crocodile-like body: its back legs were longer than its front legs. *Euparkeria* was an agile reptile that could run on four legs or two. For me, *Euparkeria* is special, not because the fossil was found in the Karoo but because it was the first reptile—except for an odd anapsid—to show signs of bipedalism, an energetically demanding form of locomotion that required an elevated metabolic rate.

FOUR

The Permo-Triassic Kick-Start

There was no worse time on earth than the Late Permian and the Early Triassic. The climate at the Permo-Triassic boundary was out of control. During the Late Permian extreme volcanic activity released vast amounts of carbon dioxide and methane gases from the South China Traps, 259.9–265.1 million years ago (Capitanian epoch), and the Siberian Traps, 254.1–259.8 million years ago (Wuchiapingian epoch), which caused extreme global warming. The warming melted the permafrost, which then released more greenhouse gases, especially methane, and added to the atmospheric heating. The process became a positive-feedback climatic horror.

Many estimates have been made of the temperature changes that occurred across the Permo-Triassic boundary. Kévin Rey from the University of Lyon and his collaborators studied the stable oxygen and carbon dioxide isotope ratios in the bone and tooth-dentine apatite of therapsids from the Karoo Basin.[1] They argue that shifts in isotope ratios over time reflect changes in the air temperature and the relative humidity of the Late Permian and Early Triassic terrestrial environments. They estimate that the average air temperature increased by about 8°C (14.4°F) 259 million years ago and by a staggering 16°C (28.8°F) immediately before the Permo-Triassic boundary, 252 million years ago. These high temperatures were sustained for about 10 million years before decreasing to those that existed before the Permo-Triassic boundary. So the Permo-Triassic was not a sudden obliteration event, such as that proposed for the extinction of dinosaurs at the end of the Cretaceous, which was caused by a meteorite's impact. According to the Rey study, the change in the climate was slow but of such sustained magnitude that it caused massive ecological deterioration that started about 13 million years before the Permo-Triassic boundary and lasted for about 10 million years into the Triassic. This was the most extreme climatic event to have occurred over the past 600 million years of the earth's history.

Yuadong Sun from the China University of Geosciences and his collaborators have also employed changes in the stable oxygen isotope ratios in conodont apatite to estimate the changes that occurred in the oceans during the Permo-Triassic boundary.[2] Conodonts are eel-like vertebrates (agnathans) that first appeared in the fossil record about 485 million years ago during the Cambrian period of the Paleozoic era. They ranged in size from 1 cm (0.39 in) to about 40 cm (15.7 in) in length and were probably plankton filter feeders. They survived the Permo-Triassic extinction and persisted until the Late Triassic. Since they lived such a long period of time in the oceans, their fossil tooth-like microstructures, called "elements," have been instrumental in defining geologic periods. Their persistence across the Permo-Triassic boundary also provides an invaluable continuous record of temperature events.

The Sun study estimated that, in the tropics, between the equator and 30 degrees north and south, the surface sea temperatures at the time exceeded 40°C (100.4°F).[2] Think about this—the sea temperature was 3°C (5.4°F) warmer than your body temperature.

In the tropics, no vertebrate could survive these temperatures, either on land or in the sea. Fish, amphibians, reptiles, and archaic mammals either died out or, if they could move, they did, to cooler, higher latitudes. The Sun study coined the term "tetrapod gap" to describe the lack of fish and therapsid fossils between 30°N and 40°S, especially during the "Late Smithian Thermal Maximum," 250.7 million years ago at the onset of the Triassic.

For marine ectotherms with high levels of activity, such as cephalopods—that is, animals such as squid and octopus—high water temperatures spelled disaster. The Arrhenius effects of elevated temperature on molecular reaction rates would have elevated oxygen demands at the same time that elevated sea temperatures would have depleted dissolved oxygen levels. These temperature effects had secondary food chain effects because they depleted the food resources of marine reptiles such as ichthyosaurs, leading to their exclusion from the tropics as well.

There is considerable debate about the actual number of organisms that went extinct around the Permo-Triassic boundary. The majority of estimates have been based on the fossil occurrences of marine organisms such as microscopic foraminifera, radiolarians, brachiopods, bivalves, cephalopods, gastropods, trilobites, and fish. Estimates range from 57 percent to as many as 83 percent of marine genera.[1] This is a staggering figure; perhaps only one out of every five genera survived the event. On land the extinctions were as dramatic. About 70 percent of all four-legged animals (tetrapods) went extinct.[1]

Although initial pulses in endothermy can be traced to the Late Permian, the Permo-Triassic boundary provided, arguably, the initial kick-start to the evolution of endothermy, particularly in the archaic mammals. I emphasize the mammal lineage at this point because the diversity and ecological dominance of the reptiles was at best marginal at that time. The reptilian opportunity was yet to come. The dramatic climate perturbations at the onset of the Triassic offered natural selection a new playground. Extreme climate changes do that. Those species that could not breed fast and persist through the changes went extinct. A small percentage did not go extinct because they possessed certain characteristics that favored their persistence. Had it not been for these characteristics, we—and all modern mammals—would not exist. So let us explore the Permo-Triassic therapsid survival kit.

The secret of the therapsids, it seems, was their *rate* of breeding, breeding *early* in life, and dying *young*.[3] These life-history attributes demanded enhanced endothermy: higher metabolic rates and body temperatures. But what evidence is there that pulses of endothermy actually occurred at the Permo-Triassic boundary?

There is currently compelling evidence, and, encouragingly, the rate at which additional new evidence is emerging in science journals is quite breathtaking.

Jennifer Botha-Brink and her colleagues reported that the patterns of bone growth in *Lystrosaurus* reveal something about their endothermic status across the Permo-Triassic boundary.[3] They have argued that the lack of growth rings and high rates of vascularization in the bones suggest that rapid, sustained growth in individuals up to the adult stage was an important factor in their success. They suggest that "exceptionally rapid growth may have contributed to the survival of *Lystrosaurus* during the end-Permian extinction and its abundance during the postextinction recovery phase."

One of Botha-Brink's collaborators, Adam Huttenlocker, then at the University of Utah, teamed up with Colleen Farmer, author of the Parental Care Model of the evolution of endothermy, and took the bone vasculature examinations one step further.[4] They applied phylogenetic ancestral character-state reconstruction techniques to estimate exactly when blood vessel passages underwent changes in size. Importantly, also, they have proposed that the size of the blood channels in the bones can be related to the size of the red blood cells that pass through them; small vessels can transport smaller red blood cells.

The advantage of smaller red blood cells is that they can take up oxygen at higher rates than larger-sized cells.[5] I would not be surprised if the ratios of surface area to volume are not constitutive of the outcome as well; relative to their volumes, small cells have large surface areas over which gas exchanges can take place. In both the synapsid and the sauropsid lineages, the ancestral members of these clades had large red blood cells. The sizes of the red blood cells of stem synapsids, such as *Sphenacodon*, for example, were estimated to have been about 10 µm (394 µin), whereas those of the modern eastern gray squirrel (*Sciurus carolinensis*) are about 5 µm (197 µin), half the size. Indeed, even in modern animals there is a link between the volume of red blood cells and the animals' metabolic rates. An ectothermic bullfrog has a red blood cell volume nearly eight times that of a human red blood cell, whereas the goat, which is a supraendotherm, has a red blood cell volume nearly five times smaller than that of humans.[5]

The Huttenlocker study revealed very clearly that the sudden miniaturization of the red blood cells in synapsids occurred at, or perhaps even slightly before, the Permo-Triassic boundary. The red blood cells continued to get smaller throughout the Mesozoic and Cenozoic and culminated in the smallest red blood cells found in modern supraendotherms, such as the rabbits and hares, and the ungulates. The same pattern can be seen in the sauropsid lineage, but a smaller sample size at present contributes to less dramatic patterns.

In modern mammals there is a correlation between the size of the heart and the size of red blood cells; mammals and birds with small red blood cells have larger hearts.[4] This relationship can be used to predict the sizes of the hearts of archaic mammals, such as *Thrinaxodon*—the "first hibernator"—which flourished in the harsh climate of the Early Triassic. *Thrinaxodon's* heart size falls within the range of those of many modern mammals, adding credence to the argument that it had elevated metabolic rates. So, apart from the probable evolution of the four-chambered heart following septation of the ventricles, the heart seems to have increased in size as well. These changes would have allowed blood to be pumped to the muscles and internal organs at faster rates and at higher systemic blood pressures, which would have allowed small archaic mammals to be considerably more aerobically active than their Permian ancestors.

If therapsids were indeed armed with an enhanced cardiovascular system, what evidence is there to support the claim that they displayed higher metabolic rates and body temperatures? How is it possible to measure the metabolic rate of an extinct animal? These questions have been tackled very recently, and the ideas that have emerged I find fascinating.

Again, one technique involves the analysis of stable oxygen isotopes ($^{18}O_p$) in bone and tooth apatite. The ratio of oxygen isotopes in bone and tooth phosphate is a direct measure of an animal's body temperature. Typically this relationship is dependent on a process known as fractionation, in which lighter isotopes of oxygen, such as ^{16}O, are favored in metabolic processes and, because metabolic processes are temperature dependent, as predicted by the Arrhenius effect, the ratio between the heavier ^{18}O and the lighter ^{16}O changes with temperature. Put simply, the heavier isotopes get "left behind" or "enriched," and their proportions relative to the lighter isotopes increase.

Kévin Rey and his collaborators, who used the same techniques to estimate air temperature across the Permo-Triassic boundary that I discussed
earlier, sampled six therapsid clades (Neotherapsida, Dicynodontoidea,
Lystrosauridae, Kannemeyeriiformes, Epicynodontia, and Eucynodontia),
whose fossils range in age from the Middle Permian, 270 million years ago,
to the Middle Triassic, 242 million years ago.[6] The data span the Permo-
Triassic boundary very nicely. These authors showed that pulses in body
temperature appeared in two clades in the Early Triassic: one in the clade
that includes the Lystrosauridae and Kannemeyeriiformes and the other in
the clade Eucynodontia, which is the clade that gave rise to the true mammals. The earliest pulses can be seen in *Lystrosaurus*, the great survivor of
the Permo-Triassic boundary that I discussed earlier, exactly at the Permo-
Triassic boundary about 252 million years ago.

It is not clear whether these two endothermic pulses evolved independently of each other in the two therapsid clades or the trait was inherited
from their common ancestor. If the trait was inherited, it would have appeared for the first time in the Middle Permian about 265 million years
ago. If not, then endothermy evolved in response to the environmental
conditions of the Early Triassic in both lineages. Endothermy went extinct
along with the Kannemeyeriiformes and the Lystrosauridae, but not in the
Eucynodontia, in which it got more and more sophisticated throughout the
Triassic and beyond, culminating in hot hares and ungulates.

It is comforting to know that sophisticated and expensive techniques
such as stable isotope analyses are not necessarily the silver bullets for peer-
ing into the past and building on the exciting new scientific discipline of
paleophysiology. Recent studies have relied on the age-old techniques of
histology to confirm that the endothermic pulses occurred at or before the
Permo-Triassic boundary and, certainly more dramatically, that it may be
possible to estimate the actual metabolic rates of the therapsids. Histology—
that is, making very thin slices of fossil bone and examining the structure
under a microscope—has been married to modern methods of phylogenetic ancestral character state reconstruction using eigenvector mapping.
Eigenvector mapping is merely a statistical method used to correlate traits
and factors.

Chloe Olivier from Sorbonne University and her collaborators estimated the metabolic rates of three dicynodontid therapsids: *Moghreberia*, *Lystrosaurus*, and *Oudenodon*.[7] Their estimates are considerably higher than measures of modern ectothermic lizards and about a quarter of those of modern endothermic birds and mammals. These species lived in the Late Permian, about 8 million years before the Permo-Triassic boundary, which suggests that the development of endothermic pulses in the therapsids was already underway before the major extinction event. These early pulses coincide perfectly with Kévin Rey's data, suggesting that the first global air temperature pulse occurred at this time.

The Olivier study did not include any Eucynodontia, so we do not know, at this point, whether this technique can confirm further pulses in endothermy in the immediate aftermath of the extinction during the Early Triassic.

The conditions of the Early Triassic tropics are highly relevant to what happened in the Karoo because the Karoo Basin was situated at around 50–60 degrees south latitude at the time, on the rim of the tropics. So it is fascinating to learn that therapsids at this time, such as *Lystrosaurus*, could have shown bursts of endothermy that allowed them to survive the mass extinction. Why might this have happened?

I suspect that the high air temperatures would have elevated the body temperatures of species of *Lystrosaurus* and conferred upon them fitness benefits such as enhanced egg development, which we discussed earlier. However, if an enhanced metabolic rate was attained through the thermal intervention of Arrhenius effects, it could have been sustained *only* if an animal had a complementary cardiovascular system that could deliver oxygen to the cells fast enough to avoid anaerobic metabolism. So it might have been at this point that the fourth chamber evolved in the heart by septation, that is, the formation of a septum that split the left and right ventricles.[8] The new ventricle allowed the blood pressure throughout the body to be considerably higher than that flowing from the heart to the lungs and back. A high blood pressure is essential for speeding up oxygen delivery to the cells.[9]

Despite the novel innovations toward endothermy that *Lystrosaurus* displayed, it was not from the dicynodonts that the true mammals evolved. The direct mammal ancestors were the cynodonts, which evolved from therocephalian therapsids. Three genera of other Therocephalia also survived the extinction event: *Tetracynodon*, *Moschorhinus*, and *Scoloposaurus* (shown in Figure A.2). What was their secret?

To answer this question we need to visualize what happened to the food chain at that time.

The ubiquitous *Glossopteris* flora had been widespread in Gondwana, and when these plants went extinct, so too did the largest therapsid herbivores and carnivores. Large therapsids were ill equipped to deal with the harsh conditions that prevailed on earth at the end of the Permian. So the simultaneous disappearance of the most dominant Permian plants, the largest herbivores, and the largest predators is hard to ignore. Again it seems that body size had much to do with who persisted and who did not in a rapidly aridifying world.

Consider *Moschorhinus kitchingi*, along with *Lystrosaurus*, another great survivor of the extinction (Figure 4.1). Fossils of this therocephalian were discovered near Nieu-Bethesda and named by Robert Broom in 1920. Adam Huttenlocker and Jennifer Botha-Brink studied the body size changes that occurred in *Moschorhinus kitchingi* across the Permo-Triassic boundary.[3] At the end of the Permian, *Moschorhinus kitchingi* was the size of a lion. However, it got smaller during the Early Triassic and also apparently grew faster, a pattern similar to that seen in *Lystrosaurus*. These data support the idea that many of the early therapsids went extinct at the onset of the Triassic because they were too big and could not breed fast enough.[3] Both predator and prey got smaller in the Early Triassic, probably because of the increasing limitation of resources. *Moschorhinus kitchingi* finally went extinct, too, because its prey species went extinct at the same time.

No members of the other major clade of therapsids, the Gorgonopsia, survived the great extinction. They were too big. One genus of gorgonopsid, incidentally, is named after the Rubidge family, *Rubidgea*.

Roger Smith from Iziko South African Museum in Cape Town and Jennifer Botha-Brink concluded that the extinctions at the Permo-Triassic

Figure 4.1. *Moschorhinus kitchingi* feeding on a *Lystrosaurus*. (Reconstruction by, courtesy of, and © Dmitry Bogdanov, http://dibgd.deviantart.com)

boundary occurred in three phases.[10] Their work offers, perhaps, the most articulate perspective of what happened at the boundary. As Roger Smith argues, the account "facilitates the documentation of details of the PTME [Permo-Triassic Mass Extinction] in this part of western Gondwana, from before onset, through the early to late phases of ecosystem collapse, into the early recovery phase, and finally to a fully developed Triassic ecosystem."

Phase one was associated with lowered water tables, which led to the "loss of shallow rooting groundcover in the more elevated proximal floodplain areas and the disappearance of the smaller groundcover-grazing herbivorous dicynodonts and their attendant small carnivores." Phase two was the main extinction event. It "reflects progressively unreliable rainfall leading to vegetation loss in proximal and distal floodplain areas. The larger tree-browsing herbivores and their attendant carnivores are confined to watercourses before finally disappearing." Phase three showed distinct evidence of aridity, including the "accumulation of mummified carcasses buried by windblown dust."

These were hard times for all organisms on earth, but especially for those in the tropics. According to the Huttenlocker study, there was a steady ecological deterioration over a period of about 200,000 years during which the food chain was weeded away slowly, and then the animals dependent on it. The large herbivores went first, followed by their large predators. The smaller herbivores and their attendant predators followed them. The last to go were the generalists, such as small omnivores and insectivores that fed on the remnant resilient insects and snails. These were conditions that fostered the evolution of endothermy in mammals—including us.

The bounce-back was immediate and dramatic. New species of *Lystrosaurus* appeared. They did not evolve as new species from the last of the survivors; they were immigrants from arid lands that surrounded the Karoo Basin before the basin dried up. They came from far afield, from the south, from Antarctica, which was still attached to Africa and Australia.

In terms of the evolution of endothermy, though, it was the dramatic evolution of the cynodonts immediately after the great extinction that demands attention. Let's start with *Procynosuchus*, one of the earliest pre-extinction cynodonts of the Late Permian and so our Late Permian ancestor (Figure 4.2).

Procynosuchus was discovered in the Karoo in 1937 and named by Robert Broom.[11] It may have been semi-aquatic, as it had a powerful long tail that was possibly used for swimming. It possessed several of the novel innovations that characterize enhanced endothermy, such as a complete braincase, which allowed for greater muscle force to be employed by the jaw muscles, a flared zygomatic arch allowing more complex jaw muscles, an almost complete secondary palate allowing breathing while eating, and dif-

Figure 4.2. Reconstruction of *Procynosuchus delaharpeae*, the earliest cynodont, which went extinct at the Permo-Triassic boundary. (Image by, courtesy of, and © Nobu Tamura, http://ntamura.deviantart.com)

ferentiated teeth, incisors, canines, premolars, and molars. *Procynosuchus* was not restricted to the Karoo; it had a wide global distribution that included other parts of Africa, namely Zambia and Tanzania, and stretched as far afield as Germany and Russia. It was from early cynodonts such as *Procynosuchus* that the Early Triassic cynodonts evolved and radiated.

Apart from the ancestor of the clade of cynodonts that gave rise to the mammals, namely the Cynognathia, four ancient cynodont genera (Epicynognathia) also bore their genes into the Triassic: *Galesaurus, Progalesaurus, Thrinaxodon*, and *Platycraniellus* (shown in Figure A.2).[12] *Thrinaxodon* (Figure 4.3) is of great interest to us because its fossils have been found in burrows in the Karoo in fully articulated curled-up positions, suggesting that they were "aestivating" when they died.[13] "Aestivation" is a term used to describe either short- or long-term torpor in response to aridity and is seldom applied to mammals; the terms "torpor" and "hibernation" are preferred.[14] Vincent Fernandez from the University of the Witwatersrand,

Figure 4.3. A scene from the Karoo Basin about 250 million years ago, in the Early Triassic, showing a *Thrinaxodon* and pups. (Reconstruction by, courtesy of, and © John Sibbick, http://www.johnsibbick.com)

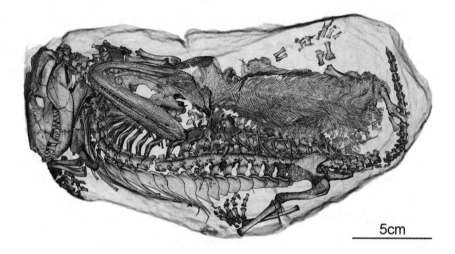

5cm

Figure 4.4. The extraordinary cohabitation of a burrow in the Karoo Basin by an injured stem amphibian, *Broomistega,* and the therapsid *Thrinaxodon,* which probably excavated the burrow. The scenario suggested is that, being in a torpid state, *Thrinaxodon* "tolerated" the injured amphibian's sharing its burrow and that flooding of the burrow killed both at the same time. The skull of *Broomistega* is resting on the shoulder of *Thrinaxodon,* whose skull is at the extreme left of the graphic. (Image from Fernandez, V., et al., "Synchrotron Reveals Early Triassic Odd Couple: Injured Amphibian and Aestivating Therapsid Share Burrow," *PLOS ONE* 8 [2013]; *PLOS ONE*, CC BY 4.0)

Johannesburg, and his collaborators used synchrotron scanning of a fossilized burrow cast to identify a *Thrinaxodon* sharing a burrow with an injured stem amphibian, *Broomistega* (Figure 4.4). (Bruce Rubidge and his collaborators named *Broomistega* in honor of Robert Broom.) The Fernandez study offers the first documented evidence of torpor or hibernation in the mammal lineage and supports my arguments for plesiomorphic (ancestral) hibernation in Chapter 14.[13] These authors argued:

> The numerous specimens of *Thrinaxodon* discovered in curled-up positions, suggest that this animal had retreated into its burrow for a period of dormancy. However, the absence of histomorphological markers indicative of arrested appositional bone growth, suggests that these periods were relatively short in duration. As torpor is viewed as a plesiomorphic character in mammals, it is

more likely that metabolic plasticity existed in this mammal fore-runner rather than the specialised metabolism of a hibernator. The capacity to escape hazardous climatic conditions in a burrow and to survive deprivation of vital resources certainly contributed to the success of small to medium-sized cynodonts across the PT [Permo-Triassic] crisis.

Using alternative terminology, *Thrinaxodon* would be called a daily het-erotherm, which means that it entered torpor on a daily basis and did not employ long-term hibernation.[14] If members of the genus had been hiber-nators, the authors further argue, their bone histology would have shown the tree-ring-like lines of arrested growth. They did not, but the verdict is still out on the paleophysiological relevance of lines of arrested growth.

These cynodonts did not persist for very long during the Early Triassic, but they persisted long enough to give rise to two major new cynodont evo-lutionary lines, the Cynognathia and the Probainognathia. It was the Pro-bainognathia that gave rise to the mammals, but, interestingly, this group did not really get going until the Late Triassic and the Early Jurassic. The Cynognathia, however, exploded during the Early Triassic after the final waning of *Lystrosaurus*. They inherited the many niches vacated by the vic-tims of the Permo-Triassic Mass Extinction, and in them evolved the adap-tations to the new conditions of the Early Triassic. But the Cynognathia also did not persist for very long. Most went extinct during the Middle Triassic, although some species, such as *Scalenodontoides*, survived until the end of this period.

So what was it about the Probainognathia that allowed them to persist into the Jurassic to give rise to true mammals?

The Early Triassic cynodonts exhibited yet more innovations toward mammalness and endothermy: an expanded zygomatic arch, differentiated teeth, a massively enlarged dentary bone in the mandible, and a complete secondary palate.

For a long time it was thought that the most advanced cynodont was a small insectivore, *Tritheledon riconoi*, named by Robert Broom during his early years as keeper of vertebrate paleontology at the South African

Museum in Cape Town.[15] However, research by Marcello Ruta from the University of Lincoln, in collaboration with Jennifer Botha-Brink and others, has shown that another cynodont family, Tritylodontidae, seems to be more closely related to the true mammals than the Tritheledontida.[12] Species of the genus *Tritylodon* looked remarkably like modern rodents. They were about hyrax sized and had large, protruding incisors and a gap in the tooth row, a diastema, where the canines would normally be seated. Their fossils have been found in both the Karoo and Antarctica.

There were very few features that separated *Tritylodon* species from true mammals. The critical non-mammal feature was the "primitiveness" of the bones in the jaw. Indeed a genus of Tritheledontidae, *Diarthrognathus*, meaning "two-joint jaw," which coexisted at the same time and in the same places as *Tritylodon* in the Karoo, takes its name from the last stage of the ancestral jaw that existed immediately prior to the development of the lower jaw in true mammals, which comprised one bone only, the dentary.

Tritylodon also possessed another new and derived dental feature: it had periodontal ligaments that anchored the first incisor to the jawbone. The presence of periodontal ligaments indicates precise occlusion between the upper and lower incisors, that is, advanced chewing ability.

However, the tritylodont that is most closely related to the Mammaliaformes was *Oligokyphus*, arguably the most derived cynodont (Figure 4.5). *Oligokyphus* becomes my generic derived cynodont character "dogtooth" in Chapter 9. It was a small cynodont, probably furry, and about 35 cm (11.8 in) long, with a slender, weasel-like body. Although the tritylodonts as a group are considered to be herbivores, *Oligokyphus* was probably more

Figure 4.5. *Oligokyphus* was a highly derived cynodont and was probably the closest ancestor of the mammaliaforms. (Reconstruction by, courtesy of, and © Nobu Tamura, http://ntamura.deviantart.com)

omnivorous and would have fed on seeds and other quality plant items that it could manipulate with its hands as modern rodents do. Like rodents, it also had a diastema, a lack of canine teeth that created a gap between the incisors and the hind teeth. The first fossils of *Oligokyphus* were found not in the Karoo but in the United Kingdom, Germany, and China.

We need to leave Africa for now because my story of the Permo-Triassic extinction and its evidence embedded in the Karoo rocks comes to an end here, right at the point where the true mammals evolved from an archaic mammal that must have been much like *Oligokyphus*. It is highly probable that *Oligokyphus* had attained an advanced level of endothermy and that this attribute would have been passed on to the mammals. All graphical reconstructions of *Oligokyphus* that I have seen show it with fur, although there is no direct fossil evidence to confirm furriness. It has been suggested that *Oligokyphus* exhibited advanced parental care behavior—looking after its young—an attribute passed on to the mammals and retained until this day. We'll pick up on these characteristics of *Oligokyphus* later, but before we do, let us catch up with what happened to the scaly, sauropsid lineage during the Permian and the Triassic.

FIVE

Reptile Takeover

W hile the mammals-to-be, the stem synapsids and the therapsids, were enjoying dominance of the earth during the Permian, on the other side of the fence their scaly neighbors, the first reptiles in the sauropsid line, were, for the most part, sprawling along. There was an early split in the reptiles between the anapsids and the diapsids (shown in Figure A.1). The diapsids are the reptiles that interest us, because they were the ancestors of the snakes, lizards, turtles, crocodiles, pterosaurs, dinosaurs, and birds. They were the reptiles that produced endotherms—the pterodactyls, dinosaurs, and birds. All anapsids went extinct at the Permo-Triassic boundary. They included reptiles such as herbivorous pareiasaurs, some of which attained a body mass of 600 kg (1,300 lb). It is generally thought that the early anapsids and diapsids were all ectotherms. It was only much later in the sauropsid lineage that endothermy evolved in the pterosaurs, dinosaurs, and birds, perhaps even as early as in the ancestor of the crocodile lineage. Indeed, the first features toward endothermy in the sauropsid line appeared in the Triassic during the great reptile takeover.

But this current wisdom might need questioning because there was one anapsid family, the Bolosauridae, which includes a 290-million-year-old (Early Permian) species, *Eudibamus cursoris*, which was a bipedal cursor that employed parasagittal locomotion. It had elongated digits in the foot and ran on its toes (digitigrade locomotion). As the authors who described this fossil—led by David Berman from Carnegie Museum of Natural His-

tory, Pittsburgh—emphasized, this small lizard-like reptile "precedes parasagittal archosaurs (including dinosaurs) by at least 60 million years."[1]

I argue later that bipedalism was probably a reasonable indicator of an
animal that had a high aerobic scope for locomotion—that is, an elevated
metabolic rate produced by endothermy. Moreover, I also argue later—in
support of Roger Seymour from the University of Adelaide—that vertebrates
that can adopt a head-up stance, such as occurs during bipedal locomotion,
require a high systemic blood pressure, which can be provided only by a
four-chambered heart. Blood pressure is positively correlated with metabolic rate in vertebrates.[2] This intriguing scenario might suggest that some
anapsids had attained endothermy long before any other lineage that went
on to produce crown endothermic birds and mammals. So endothermy
may have evolved twice independently in the sauropsids, once in the anapsids, and again in the diapsids, but this is not reflected in modern sauropsids because all anapsids went extinct at the Permo-Triassic boundary. On
the other hand, perhaps *Eudibamus cursoris* was something like the "Jesus
Christ lizard," *Basiliscus basiliscus*, a modern ectothermic lizard that runs
on its hind legs on water to escape predators, but for short periods of time
only. Hoard this fascinating prospect for later when I suggest that the timing of the origins of endothermy may be much older than we currently
appreciate.

An important split in the diapsid lineage occurred 240 million years ago,
during the Middle Triassic; the Lepidosauria branched off to give rise to the
snakes, lizards, and turtles. Lepidosaurs remain ectothermic to this day, so
we can be confident that ectothermy was still the prevailing physiological
state when the split took place. Or can we?

Burmese pythons and tegu lizards, which I discuss later, can produce
endogenous heat when breeding, so the lineage is not strictly ectothermic
throughout. The ancestor of the lepidosaurs may indeed have had a rudimentary capacity to produce endogenous heat that was retained in a few
large-bodied snakes and lizards. Alternatively, the production of muscular
heat could have evolved independently in large-bodied lepidosaurs.

The emergence of the Lepidosauria is owed to the Permo-Triassic Mass
Extinction, as is that of the group from which they split, the lineage leading

to the Archosauria, which means "ruling reptiles." The Archosauria radiated rapidly and went on to numerically and ecologically dominate the archaic mammals during the Late Triassic, the Jurassic, and the Cretaceous (shown in Figure A.1). It was the archosauriforms and the more derived Archosauria that staged the great reptile takeover. The archosaurs include the living crocodiles, the extinct nonavian dinosaurs and pterosaurs, and the birds. The early archosaurs took over the niches once occupied by the carnivorous therapsids—gorgonopsids and titanosuchids—that were decimated by the Permo-Triassic extinction (discussed in Chapter 3). The archosaurs went on to replace the larger-bodied therapsids and cynodonts by the Late Triassic. The notable radiation of the archosaurs was the split into the two lineages that have living representatives today, the Pseudosuchia (crocodiles) and Avemetatarsalia (birds).

Why and how the archosaurs were able to achieve the Triassic takeover is a subject of great interest to paleontologists. It is also of interest to us with respect to the evolution of endothermy because physiological as well as anatomical characteristics are implicated.

During the Early Triassic there was only one continent on earth, the giant Pangaea. Pangaea started to split apart in the Middle Triassic into the northern continent Laurasia and the southern continent Gondwana. But prior to the split, the climate of the Early Triassic was vastly different from that of the Permian. It was hot and dry and, for the most part, desert-like. The polar regions, though, were more moist and temperate. The dramatic aridification following the Permian is thought to have been the principal explanation for why the archosaurs were able to supplant the therapsids and the cynodonts, such as the ubiquitous dicynodont *Lystrosaurus*, which survived the Permo-Triassic extinction.

The Early Triassic archosauriforms were the first to exhibit the thecodont condition, which is one of the defining features of the Archosauria. "Thecodont" means that the teeth are set in sockets ("socket teeth") that anchor them and give them stability during feeding. So here we go again, encountering the same old story, just as in the synapsid mammal lineage: improved tooth arrangements can be regarded as one of the first innovations toward potential endothermy in sauropsids, because they allowed food to be processed more efficiently.

Another defining archosaur characteristic was the appearance of an opening in the skull in front of the eye socket, the preorbital fenestra, which lightened the skull and again improved feeding efficiency. The linings of the hole in the bone also provided new surfaces for jaw muscle attachments. The weight of the mandible was lessened with the evolution of another hole, the lateral mandibular fenestra, an opening in the side of the jawbone.

Later, in some bipedal archosaurs, another new critical innovation evolved—a fourth trochanter—a bony ridge on the back side of the femur to which additional muscles could be attached, which allowed the animals to walk upright on two legs. Some archosauriforms were on their way to being bipedal, a condition that is seen today in birds and, as we shall see, was paramount for the evolution of endothermy in dinosaurs and birds.

The prevailing wisdom has it that archosauriforms were ectothermic, mostly, and had a body covering of scales, we think. They certainly had a waterproof skin provided by the lipids in the skin epidermis. They also produced uric acid instead of urea as their primary means of getting rid of nitrogenous wastes. This feature was to stand them in great stead as the desert-like Early Triassic unfolded, because it gave them massive water-conserving benefits over the therapsids and cynodonts.

Large ectothermic reptiles excel in deserts. During the Pleistocene, for example, when the conditions in Australia were similar to those of the Triassic, the largest terrestrial predators were enormous crocodiles and lizards, not carnivorous marsupial mammals, as one might have expected given the complete dominance of the Australian endotherm fauna by marsupials today. These megafaunal ectotherms disappeared between 40,000 and 30,000 years ago, probably driven to extinction by the arrival of *Homo sapiens* in Australia. *Varanus priscus* was a giant monitor lizard that reached a length of 7 m (23 ft) and weighed nearly 620 kg (1,367 lb). It looked much like the Komodo dragon, to which it is related, but it was considerably larger—nearly ten times larger. Like the Komodo dragon, though, it had toxin-secreting oral glands in its mouth that would have made it the largest venomous vertebrate that ever existed. *V. priscus* roamed the deserts of Australia side-by-side with another ectothermic monster, a completely land-based crocodile, *Quinkana fortirostrum*, which reached lengths of 6 m (20 ft). The only mammalian carnivore that could remotely compete with these ectothermic giants was the marsupial lion, *Thylacoleo carnifex*, which weighed a mere 150 kg (331 lb) or so, considerably less than today's African lion. Reptiles love deserts.

In Chapter 2 I discussed how the evolution of uric acid as a nontoxic waste deposit paved the way for the success of the cleidoic egg and the colonization of land from the swamps. In the sauropsid lineage, adults also excreted uric acid as the main product of waste nitrogen. They could do this because they had a cloaca, a single posterior opening into which the intestinal, urinary, and reproductive tracts open. It is in the cloaca that the uric acid precipitates out as a white waste from the excretions of the urinary tract. The nitrogen, of course, comes from metabolized amino acids and proteins. The massive advantage of producing uric acid as adults compared with producing urea, which mammals do, has nothing to do with the toxicity issue. It concerns problems with water. To excrete urea, a certain amount of water must be added to the urea to keep it sufficiently dilute to prevent it from becoming toxic in the body. The kidneys, depending on the hydration status of the animal, control the amount of water very well, but, even in the most dehydrated state, mammals use a lot of water to excrete urea. In animals that produce uric acid, very little water is required for its excretion. This is the principal reason why birds and reptiles do so well in deserts. They can get by on much less water per day than mammals can. This may be the principal explanation for the reptile takeover of the therapsids after the Permian.

I have watched with a mixture of amusement and admiration footage from the popular documentary series *Walking with Monsters* showing a Day-Glo green *Euparkeria* (introduced in Chapter 3) darting around on its hind legs and leaping into the air to snatch a passing dragonfly-looking insect. The documentary hints that *Euparkeria* may have been an extremely agile archosauriform, possibly the first to be so twinkle-toed, and that this capacity was passed on to later dinosaurs that became fully bipedal. It's a fun suggestion — I love the passion, imagination, and knowledge of the reconstructors of these long-gone reptiles — but how did those who created *Euparkeria*'s antics know that she had the aerobic metabolism to carry out such feats of

athleticism? Some modern lizards can display some very athletic perfor-
mances, such as the Jesus Christ lizard *Basiliscus* from South America that
I have mentioned. But for *Euparkeria* to have been able to do what I saw
her doing in *Walking with Monsters*, she would have needed the endotherm
cardiovascular toolkit to provide large amounts of oxygen to her muscles.
Euparkeria would have needed a four-chambered heart and a high blood
pressure to catch her "dragonfly" with such elegance. The makers of *Walking with Monsters* cannot have known that *Euparkeria* had an endotherm's
heart. Hearts do not fossilize.

But who knows? Maybe *Euparkeria* did indeed have a four-chambered
heart.

For simplicity we can say that modern lizards and snakes—with some exceptions—have three-chambered hearts that make it impossible to separate
the blood pressure between the heart and the lungs from that between the
heart and the main circulation of the rest of the body. In a four-chambered
heart, such as ours, the blood pressure throughout the body, the systemic
blood pressure, is much higher than in the lungs, the pulmonary pressure.
This is because of the strength and power of the fourth heart chamber, the
left ventricle, which pumps the blood into the circulation system. (Blood
pressure is measured normally as the systolic pressure, or maximum arterial
blood pressure during one heartbeat, relative to the diastolic pressure, the
minimum arterial pressure between heartbeats. The averge for humans is
120/80, systolic over diastolic. The systemic pressure is a rough average of
these two measures.) A four-chambered heart and a high systemic blood
pressure are essential for providing a high sustained oxygen supply to an
endothermic body—that is, a body that can indulge in sustained activity.
My average systemic blood pressure is about 110 mm Hg when I don't have
to deal with idiots. That of a lizard is about 35 mm Hg, or around a third
of mine. But our pulmonary blood pressures are about the same, roughly
25 mm Hg, a quarter of my systemic blood pressure. These are the differences and similarities between the ectotherm three-chambered heart and
the endotherm four-chambered heart.

A high blood pressure is also essential for an animal that maintains an
upright stance because the blood needs to be pumped to the regions of the

trunk that are higher than the heart.[2] That is why the giraffe has the highest blood pressure of all modern mammals—there is quite an altitude change between its heart and its head. And indeed, simply because large terrestrial animals hold their heads higher above their hearts than small animals do, a relationship exists between blood pressure and body size: blood pressure increases with body mass.[2] That is why Roger Seymour and his colleagues have emphasized that large-bodied sauropsids and synapsids were probably endothermic.[3] This claim has enormous implications in terms of pinpointing the timing of the first emergence of endothermy in the family trees of the sauropsids and synapsids, but it is perplexing, to me anyway, that the idea has not been explored more extensively.

So what would have been the required arterial blood pressure of a giant sauropod dinosaur rearing up on its hind legs and stretching its neck as high as it could to reach treetop vegetation? Seymour has calculated that the sauropod *Barosaurus*, which had a body mass of 20,000 kg (44,092 lb), would have required a blood pressure of 750 mm Hg and a left ventricle with a mass of more than a metric ton (1 tonne) to achieve this feat. He argues that this was impossible, a claim he made in the prestigious journal *Nature* in 1976 that has been virtually ignored by sauropod paleontologists from North America and Europe.[4] As of February 2018, his article has been cited twenty times according to the ISI (Institute for Scientific Information) Web of Science. So the impact factor is about 0.5—that is, a citation once every two years. The typical impact factor for *Nature* articles is 40, which means that the average publication in *Nature* is cited 40 times per year. As I pointed out concerning Raymond Dart's work on hominids, and a tendency that I identify again later in this book concerning my own work, it is not atypical that the ideas of southern hemisphere biologists get smothered by the sheer magnitude of opinion of northern hemisphere biologists.

I visited the Berlin Museum of Natural History in September 2015 and found that the entrance hall has a mounted skeleton of *Brachiosaurus*, now called *Giraffatitan*. For a long time this sauropod was believed to have been the largest animal ever to walk the earth. Its size is staggering. Some specimens could have exceeded 50 metric tons (110,231 pounds).[5] The tiny pinprick head, smaller than any of its feet, is perched on a long neck subtended by a supremely long tail, and it looks quite ridiculous. The head hovers about a meter below the rafters of the roof, and the entrance hall of the museum is huge. Fortunately there is a model of the skull on the floor, because you'd need binoculars to look at the mounted one.

Seymour claims that the giant sauropods mounted in museums, with their heads held high, are complete misconceptions of the capabilities of giant sauropods. But, according to him, the ruse did stop the Berlin Museum of Natural History from raising the height of the head from 11.87 m (40 ft) to 13 m (43 ft), at a cost of about 18 million euros. Seymour suggests that giant sauropods were aquatic, could float well, and used their long necks to feed off vegetation on the bottoms of water masses.

Sauropods walked on four legs. What about bipedal dinosaurs that walked on two legs, in an upright stance, like all of the theropod dinosaurs in the bird lineage?

No bipedal animal can run fast, upright, for any period of time without a high systemic blood pressure. That is why there are no bipedal ectotherms, with the possible exception of that curious anapsid, *Eudibamus cursoris*, which I mentioned earlier. Bipedal dinosaurs must have had the high metabolic rates that are coupled to high blood pressures. Bipedal dinosaurs must have been endothermic to various degrees.

How did a four-chambered heart suddenly appear somewhere in the reptile lineage?

Well, it evolved when the single left ventricle of the ancestral heart "septated"—formed a septum, or separation—that split the ventricle into two chambers.[6]

Clues about whether *Euparkeria* might have already had a four-chambered heart can be provided by characteristics of the modern crocodile. *Euparkeria* was the proto-archosaur from which the crocodile lineage, the Crurotarsi, evolved. It is therefore the closest noncrocodilian to the crocodilians. Let's say that it was the ancestral crocodile.

Roger Seymour and his collaborators, including Gordon Grigg from the University of Queensland, have argued that the "stem archosaurs," which include *Euparkeria*, had a four-chambered heart and that this heart still exists today in modern crocodiles.[3] In other words, they are suggesting that endothermy existed in the proto-archosaur, in which case it occurred in *Euparkeria* and would have been passed on to pterosaurs, dinosaurs, and birds.

Do not imagine that the crocodile lineage was characterized by what we see in modern crocodiles and alligators, namely aquatic, ectothermic "sit-and-wait" predators. The truth is far from this. During the 247 million years from *Euparkeria* to the modern crocodiles, there were land-based terrestrial predators—active, land-conquering hunters—as well as aquatic species. They were not necessarily bipedal (some were); this characteristic was mostly retained in the noncrocodilian sister clade that gave rise to the dinosaurs and the birds.

An interesting example of such an early terrestrial crocodylomorph is *Terrestrisuchus* ("land crocodile") (Figure 5.1). The skeletal reconstruction, based upon a fossil excavated in Wales,[7] shows a slender lizard-like ancestral crocodile that was completely terrestrial and could run fast on four perfectly parasagittal, long legs positioned under the body that could work in unison to gallop. *Terrestrisuchus* weighed about 15 kg (33 lb), the size of an overweight miniature dachshund. It had a tail that was longer than its body, which probably acted as a counterbalance to allow it to adopt a more upright stance and to run on two legs when needed. It is extremely unlikely that an animal with these characteristics would have been ectothermic.

By the Cretaceous, though, there was a shift from comparatively small-bodied cursorial, terrestrial crocodyliforms to larger-bodied crocodiles that

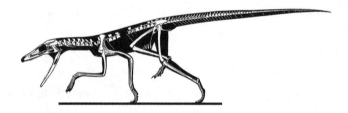

Figure 5.1. Reconstruction of the slender skeleton of the land-based crocodylomorph *Terrestrisuchus* (overall length 80 cm, or 31.5 in). Note the hint of digitigrade locomotion, walking on the toes, made possible by the elongation of the foot bones, the metacarpals in the front foot, and the metatarsals in the hind foot. (Image by, courtesy of, and © Jaime A. Headden, http://qilong.deviantart.com, CC BY 3.0)

abandoned land for the aquatic niche. Remarkably, they reverted from endothermic, widely foraging hunters to ectothermic, sit-and-wait, aquatic hunters like their ancestral, swamp-dwelling Carboniferous stem amniote ancestors. So the four-chambered crocodile heart is most likely to have been a ghost relict of an endothermic past.

Modern baby crocodiles might behave a bit like *Terrestrisuchus*, according to the Seymour team's argument: "The jumping, insectivorous habits of juvenile crocodilians give way to sit-and-wait habits of the adults . . . [and] these ontogenetic shifts in form and behavior mirror the phylogenetic shift from terrestrial to aquatic behavior of crocodilians. That adult crocodilians had to retain some terrestrial adaptations can be explained by their need to lay their cleidoic eggs on land. Water in the pores of the eggshell suffocates the embryos."

If we are to peg the origin of endothermy in the reptile lineage to a particular group, I'd be tempted to start with *Euparkeria*.

I find it quaintly amusing that I should be arguing here that Robert Broom's *Euparkeria*, a reptile that preceded the dinosaurs, could have been an endotherm. When I was a graduate student, this idea would have been perceived as preposterous. Indeed, in those days it would have been laughable to suggest even that birds were dinosaurs. So much new information has emerged over the past few decades, especially from the stunning fossils from China, that it is hard to keep track of the novelties that are pouring out week after week about the origin of dinosaurs and birds.

I'd start with an endothermic *Euparkeria* because it is the first diapsid reptile to show evidence of cursoriality, the ability to run fast. Endothermy did not necessarily evolve in birds for the same reasons that it evolved in mammals. The two evolutionary appearances of endothermy were highly independent of each other despite both lineages' having shown uncannily similar patterns of body size miniaturization, which I discuss later. In mammals the driving force for the production of internal heat, I'd argue, was the capacity to become active at night, once it had cooled, not to go running but simply to do what it takes to reproduce (discussed in Chapter 9). This idea was first proposed in 1976 by Alfred Crompton and colleagues from Harvard University in Massachusetts.[8] These activities would have involved foraging, territorial defense, courtships, incubation of eggs, and caring for the young. But none of these activities required the mammal to be a cursor. Indeed, it is only in the Paleocene, long after the dinosaurs went extinct 66 million years ago, that we see the evolution of cursorial limb forms for the first time in the mammal lineage.[9] During the nocturnal realm of the early mammals, the archosaurs and the dinosaurs "prevented" archaic mammals from becoming cursors. But once the mammals inherited the earth, it was a very different story. The mammalian carnivores and the large herbivores became fast—some very fast—as fast as a pronghorn and a cheetah.

There may be a flaw in this argument, because it has been suggested that *Euparkeria* was nocturnal and had large eyes for better nighttime vision.[10] However, I am a bit uneasy with this innovative application of the modern comparative method, because the analyses cannot easily control for running speed. Fast mammals, for example, follow Leuckart's Law and have big eyes.[11] So I'd ask, Did *Euparkeria* have big eyes because it was nocturnal, or did it have big eyes because it needed the visual acuity to dash after passing insects?

Following the appearance of *Euparkeria*, the archosaurs underwent a major split into the Crurotarsi, the crocodile lineage, and the Ornithodira, which includes the pterodactyls, dinosaurs, and birds. It was within the ornithodiran lineage that endothermy was retained and expanded upon, as can be seen in the crown group today, the birds. The ornithodirans split later into the Pterosauria and the Dinosauria. However, before that split took place, another very curious archosaur appeared, *Scleromochlus taylori*, a reptile the size of a starling (Figure 5.2).

There is evidence that *Scleromochlus* was a highly agile cursor. It may even have been a hopper, much like modern rodent jerboas or the thrushes in my garden. It was one of the first diapsid reptiles to possess a digitigrade foot, that is, a foot in which the bones between the ankle bones and the toe bones, the metatarsals, are exaggerated in length, reduced in number from five to four, and compacted into a single functional bundle. Digitigrade feet occur in fast runners, like modern mammalian carnivores. They allow their bearers to run on their toes, the digits, and avoid flatfoot heel striking. The evolution of the digitigrade foot allowed the entire hind limb to

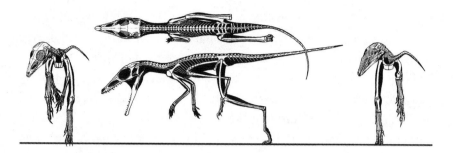

Figure 5.2. *Scleromochlus taylori* (overall length 18 cm, or 7 in), a Late Triassic archosaur that displayed some of the first indelible indicators of an endothermic capacity—it could run fast—it was a cursor. It had especially long, slender hind legs and elongated hind foot bones (metatarsals), accompanied by a reduction in the relative size of the front limbs, a characteristic that was retained in all dinosaurs leading up to modern birds. (Image by, courtesy of, and © Jaime A. Headden, http://qilong.deviantart.com, CC BY 3.0)

become longer, which increased the stride length and hence the speed of the animal.

Euparkeria and *Scleromochlus* are very thought-provoking reptiles. They were smallish, slender, cursorial predators, and they probably had four-chambered hearts, one of the hallmarks of an endotherm. Their hearts would certainly have given them the capacity for aerobic metabolism, but to make efficient aerobic metabolism possible, they needed other innovations: increased mitochondrial density in the muscles to process ADP into ATP, a system of breathing that allowed simultaneous running and breathing, and a means of producing heat.

Andrew Clarke from the British Antarctic Survey, Cambridge, and Hans-Otto Pörtner from the Alfred Wegener Institute in Bremerhaven, Germany, have argued that sustained running requires warm muscles—that heat is essential for muscle power.[12] For sustained aerobic respiration during running, ATP must be generated rapidly to fuel the energy demands of the muscles. These authors show that ATP generation by the mitochondria is strongly dependent on temperature. Warm muscles are necessary to sustain fast locomotion. No creature exists on earth today that can run at high speeds for long periods of time with a low body temperature—let's say, lower than 30°C (86°F). *Euparkeria, Terrestrisuchus,* and *Scleromochlus* had long, slender hind limbs—the attributes of a cursorial animal—suggesting that there was strong selection for the cursorial limb at this time. I will discuss the merits of this warm muscles idea more in the final chapter.

Robert Bakker was perhaps the first paleontologist to recognize that bipedal dinosaurs that could run fast needed to be endothermic. He published an insightful paper in *Nature* in 1972, titled "Anatomical and Ecological Evidence of Endothermy in Dinosaurs," in which he argued that "some small dinosaurs . . . probably could reach top speeds similar to those in ostriches and ungulates, that is, 50–80 km h^{-1} [31–50 mi/h]. . . . Assuming the presence of lizard-like energetics, the maximum speed of a 100 kg [200-lb] ostrich dinosaur would be only 2.9 km h^{-1} [1.8 mi/h], and that of a 10 metric tons [22,046-lb] tyrannosaur only 5.8 km h^{-1} [3.6 mi/h]. Clearly, dinosaurs were built for sustained speeds much higher than these, and energy metabolism in these creatures must have been like that of endotherms."[13]

Remember, Bakker wrote this in 1972, long before we knew that birds are indeed dinosaurs and long before the Chinese scientists finally revealed to the world their truly outstanding fossils of feathered dinosaurs. Robert Bakker did not merely make a blind guess. He actually measured the running speeds of living reptiles of different sizes and from these data could make an educated guess by upscaling his data to larger body sizes. He used the comparative method, the same method used by Darwin to help him understand the links between the form and function of extinct and extant animals.

Unless these early reptilian cursors had some form of insulation, they would not have been able to maintain a constant warm body temperature through increased metabolism, as we see in insulated endotherms today. But so what if they couldn't? Provided that they were diurnal hunters, they could have used normal basking behavior like that we see in modern lizards to increase their body temperature. Once the preferred body temperature had been attained, they could go hunting and rely on their recently evolved cardiovascular, mitochondrial, and breathing innovations to provide sufficient oxygen for sustained aerobic capacities. Endotherms do not need to be the typical insulated thermoregulators, as we know them today, to enjoy the fitness benefits of aerobic capacity. There is no reason to presume that the same might not have applied to some therapsids as well. Of course, how this metabolic scenario might have been reflected through lines of arrested growth in the bones is not certain, especially without considerations of seasonal metabolic slow-downs.

Ideas about the evolution of bipedalism in the bird lineage have been rocked somewhat by a new hypothesis suggesting that the large thigh muscles of bipedal dinosaurs and birds were also heat generators. Stuart Newman and his colleagues from the University of Arizona in Tucson, Arizona, have proposed the Thermogenic Muscle Hypothesis (TMH), in which they argue that archosauriforms and nonavian archosaurs remained endothermic by "selection for an enhanced capability of skeletal muscles to generate heat."[14] These authors have proposed that muscular heat production occurred through biochemical adaptations (discussed in Appendix 1) and appreciable increases in the mass of skeletal muscles. "Hyperplasic thigh muscles changed the anatomy of the evolving proto-avians," they claim, "inclining them toward bipedality." With the focus of locomotion and heat production now on hefty thighs, the upper limbs were open for selection for novel functions such as swimming, flying, titillating mates, climbing, scratching, catching things, or nothing. Selection for hind-limb muscle mass led to the evolution of bipedalism—an ability to walk on two legs— that has been retained by birds to this day.

A muscular source of heat, which presumably still occurs in birds today, is heat that is produced over and above the heat from the basal metabolic rate produced by leaky membranes in the internal organs. It is heat generated on demand, and its source is very different from the non-internal-organ heat sources in mammals. Consequently, compared with mammals, birds have a much greater muscle mass in proportion to their total mass. That is why I choose to eat a chicken over a similar-sized rabbit; they taste the same, sort of, but there is much more meat on a chicken thigh, my favorite bit of the hen.

I suspect that *Euparkeria* and *Scleromochlus* exhibit the first evidence of endothermy in the lineage leading toward birds simply because they were the first to show that the hind limbs were adapted for fast bipedal locomotion. Although the TMH provides a possible explanation of *how* endothermy evolved in birds and dinosaurs, it does not offer any suggestions about *why* endothermy evolved. My argument would be that, in addition to the benefits of endothermy per se, hot muscles facilitated fast, *sustained* locomotion. Initially, the evolutionary process might have operated like positive feedback—more muscle, more heat, more speed, more distance. Not enough? More muscle, more heat . . . and so on, until maximum possible thigh muscle mass was attained. Trade-offs would have put a ceiling on thigh muscle mass, because at some stage it would have impeded locomotion efficiency, with thighs too huge to run fast. But selection for enhanced locomotion may also have been a secondary spin-off of selection for heat-producing muscles. It may have been a secondary consequence of the primary selective pressure, such as speeding up egg development time. These dichotomous prospects emphasize Tom Kemp's argument that endothermy was unlikely to have evolved through a single-cause focus, discussed in more detail in Chapter 18.

Whether *Euparkeria* and *Scleromochlus* were endothermic or not will, for at least the time being, remain debatable, but we can be more certain of the next step toward endothermy, seen in the closest relative to the members of *Scleromochlus*, the pterosaurs. Not only were they endotherms; they were the first and biggest endotherms to fly. Before we get airborne, though, I need to introduce two cornerstone concepts associated with the evolution of endothermy: body size miniaturization and insulation.

SIX

The Great Shrinking

In the derived cynodonts and the nonavian theropod bipedal dinosaurs, including the proto-dinosaur *Saltopus* (shown in Figure A.4), a profoundly important new trend, certainly in terms of the evolution of endothermy, took hold. The ancestors of birds and mammals started to shrink, dramatically—they got smaller. Miniaturization commenced at more or less the same time, during the late Middle Triassic.

Brian McNab from the University of Florida wrote an eloquent paper in 1978.[1] It got smothered by later hypotheses on the evolution of endothermy, so it has never really been given due approbation. I have seen it referred to in various publications as the "Miniaturization Hypothesis," which is what I, too, shall call it.

As an animal physiologist, McNab was the first person to articulate the importance of body size changes during the evolution of endothermy. I am not talking here about the same acute early insights that Robert Bakker advanced for endothermy in dinosaurs six years earlier.[2] McNab was focused on the mammalian lineage. He emphasized how the lineage leading to mammals changed in body size over 100 million years, from small sizes in the earliest haul-out stem amphibians through large body sizes in the stem synapsids and early therapsids and back again to small body sizes in the advanced cynodonts. At the time, the really tiny Mammaliaformes, such as the 2-g (0.07-oz) *Hadrocodium*,[3] had yet to be discovered and described.

McNab would have had a field day if he could have strengthened his argument by adding this diminutive synapsid to his data set.

After the cynodonts, mammals stayed small—by small I mean less than a kilogram (2.2 lb), with a few exceptions I discuss later—for the next 140 million years, until the demise of the dinosaurs 66 million years ago. Only then did they radiate into the spectacular diversity of sizes that we see today, which ranges from a 2-g (0.07-oz) bat to the 180,000-kg (396,832-lb) blue whale. This Cenozoic increase in body size from small endothermic mammalian ancestors at the end of Cretaceous could never have happened had miniaturization not occurred in the Triassic and Jurassic, because it was reliant solely upon endothermy in a rapidly cooling world.

Why should body size changes have influenced the evolution of endothermy?

Brian McNab argued that the answer to this question concerns the preservation of the fitness benefits of a high thermal inertia in ancestral forms that had started to show enhanced levels of endothermy.

Permit me to explain thermal inertia in familiar terms. My swimming pool cannot cool by more than 1°C (1.8°F) overnight, no matter what the weather throws at it; it contains 80,000 liters (21,133 US gal) of water, equivalent, more or less, to 80,000 kg (176,369 lb), which is nearly half the size of a blue whale. But my chilled glass of Chenin Blanc can warm up in the same environment from an ideal 5°C (41°F) to a ghastly 25°C (77°F) in less than an hour. The difference is attributed to thermal inertia, which involves surface-area-to-volume ratios; large bodies have a much smaller surface area relative to their volume over which they can lose and gain heat. My swimming pool behaves like an endotherm in one sense; it remains at a constant temperature. Of course it is not endothermic because it cannot produce its own internal heat. The term used to describe its thermal properties is "homeothermic," which means maintaining a fairly constant temperature.

Brian McNab advanced the idea that large stem synapsids and early therapsids were homeothermic by virtue of the large body sizes that they attained during the Permian and their consequent high thermal inertia. It is the same idea that Bakker proposed for large dinosaurs. The body temperatures of early therapsids might not have been equivalent—certainly not as elevated—to those of modern mammals, but their temperatures did not fluctuate to the great extent to which, for example, a modern 10-g (0.35-oz) lizard's body temperature would fluctuate on a daily basis.

Data for the modern 70-kg (154-lb) Komodo dragon (*Varanus komodoensis*) illustrate the concept quite nicely. The Komodo dragon is the largest living lizard. It lives in the tropics on a few islands in Indonesia that form the eastern end of the Malay Archipelago. The temperature conditions in which it lives are cooler than those of the Early Triassic, yet the body temperature of this lizard fluctuates by a mere 6°C (10.8°F) on a daily basis, and it is able to attain and retain a preferred body temperature of 34.0–35.6°C (93.2–96.0°F).[4] The Komodo dragon can do this without having to produce its own heat. This homeothermic capacity—the relative constancy of its temperature—can be attributed solely to its large body size and high thermal inertia. A 10-g (0.35-oz) lizard cannot do this; its body temperature would remain more or less the same as that of the ambient air except when it is actively basking in the sun. Large-bodied homeothermy means not that large reptiles necessarily achieved higher body temperatures than small ones, but merely that the daily fluctuations in their body temperatures were smaller, irrespective of the average air temperature.

McNab did not speculate on what the fitness advantages of being a homeotherm during the Permian might have been. There are several possibilities, and they all involve enhancement: activity levels, foraging efficiency, digestive efficiency, growth rate, egg development, and so on. Again, enhanced performance would have been related to the relative constancy of the body temperatures rather than to the absolute body temperature attained at one time during the day. These fitness benefits would have been made possible mostly because enzyme systems could operate within a narrower range of temperatures and thereby function optimally. I like this idea, because once these fitness benefits had been enjoyed by these largish homeotherms, there would have been strong selection to *preserve* them, which helps to explain subsequent events in the drive toward endothermy.

At this point a fascinating twist occurred in the story of the evolution of endothermy, certainly in the synapsids' lineage. The synapsids started to get small—very small by the standards of Permian synapsids. If the early therapsids were enjoying fitness benefits from their largish body sizes through homeothermy, then why did they start to get smaller toward the end of the Permian and Early Triassic?

McNab suggests that ecological factors were involved: "It can be concluded that the evolution of a small body size in carnivorous therapsids was associated with, or produced by, a shift in food habits, at first from large to small vertebrates and, as the decrease continued, from small vertebrates to invertebrates, especially insects. This shift in prey selection may have issued from competition with the number of large, terrestrial carnivores . . . and from a paucity of predators on small vertebrates and on invertebrates."

However, current wisdom argues that that miniaturization was not a consequence of competition with sauropsid reptiles. I reserve further discussion on the potential drivers of body size miniaturization for the end of this chapter and for the last chapter. But another of McNab's ideas concerning the physiological implications of a sudden size reduction in therapsids is certainly worthy of debate at this point because I think that it is a profoundly important idea.

McNab had an intuitive grasp of allometric scaling in animal physiology—the relationship between body size and physiological characteristics—which is essential for understanding the implications of miniaturization. I don't want to introduce concepts that are unfamiliar to a nonspecialist, so permit me to simplify the terminology slightly to explain part of McNab's argument and also an important concept in animal physiology: thermal conductance. This term refers to the ease with which heat can enter or leave an animal's body. Animals with a low level of conductance lose and gain heat more slowly than do those with a high level of conductance. The value of conductance is determined by a multitude of factors: insulation (type and thickness), peripheral blood flow patterns, posture, and especially body size. Thus, although insulation is a single component of an animal's thermal conductance, it is obvious that an animal with a high level of insulation will lose and gain heat more slowly than one with a low level of insulation. But here's the critical point: irrespective of the nature of the insulating medium, be it fur, blubber, or feathers, given the same type of insulation, smaller animals have a higher level of thermal conductance than do bigger animals.

This means that they are more susceptible to losing heat to the environment when it is cold, or gaining heat when it is hot.

As we have seen, endotherms need to defend their body temperature against heat losses by producing heat internally to match the heat that is being lost. So because small animals have a higher conductance, they lose heat faster relative to their size than do larger animals and hence need to produce more heat per unit of time and mass than do larger animals.

In general, ectotherms have poorer insulation than endotherms because they do not possess a means of trapping air on the skin surface. For example, lizards have scales that do not trap air. So here's the oddity: whereas small ectotherms do indeed have poorer insulation than do similar-sized endotherms, Brian McNab identified that the difference that this leads to in their conductance starts to diminish as body size increases to the point where a 30–100-kg (66–220-lb) ectotherm has about the same level of conductance as does a similar-sized endotherm.

The interesting idea that McNab introduced was that to retain the fitness benefits of homeothermy, miniaturization *must* have been accompanied by selection for higher rates of respiration per unit mass and time, which involves, quite simply, a pulse in endothermy: "The only way that the thermal independence of an inertial homoiotherm can be transferred to an animal with a small mass is to convert from ectothermy to endothermy."

For example, a 20-kg (44-lb) hypothetical ectothermic homeotherm could become a 2.2-kg (4.9-lb) endothermic homeotherm without any extra increase in the total amount of energy consumed. If McNab is right, it means that the first physiological machinery that allowed animals to produce internal heat on demand, specifically for the purpose of balancing heat that gets lost from the body, evolved first during the miniaturization process. It would have evolved in the smallest cynodonts leading to the mammals, and, as I will argue later, the same principal would have applied equally to the smallest theropod dinosaurs and some of their more immediate ancestors.

One year after McNab's paper appeared, Albert "Al" Bennett and John Ruben published one on the smotherer: a new hypothesis on the evolution of endothermy that became known as the Aerobic Capacity Model.[5] The hypothesis became the most controversial and popular in the whole endothermy debate and kick-started the modern dispute about the evolution of endothermy.

The Aerobic Capacity Model argued that selection for increased metabolic rate in the stem synapsids and therapsids allowed them to increase their capacity for activity, for movement, and for sustained locomotion. Bennett and Ruben do not believe that there was selection for the basal metabolic rate. They also do not believe that selection had anything to do with trying to produce internal heat so that the animal could maintain homeothermy or the elevated, regulated body temperature that we see in modern endotherms. In other words, they argued that thermoregulation was not the initial innovation that led to endothermy in birds and mammals. They suggested what would have been the fitness advantages of an enhanced aerobic capacity in an early endotherm: "An animal with greater stamina has an advantage that is readily comprehensible in selective terms. It can sustain greater levels of pursuit or flight in gathering food or avoiding becoming food. It will be superior in territorial defense or invasion. It will be more successful in courtship and mating. These advantages appear to us to be worth increased energetic costs, particularly since the enhanced capacities give their possessor the ability to increase energy intake to meet new energy demands."

These are seductive ideas, and attempts to collect data to test them have been wide-ranging. The tests have met with mixed success. It is a difficult hypothesis to test, and I discuss the problems in more detail later. Pigeon-hole the ideas for now, because they become important in later dialogue when I will discuss the major alternative, direct selection for an elevated, constant body temperature.

Apart from the first emergence of hair, there are other features of stem synapsids and therapsids that have been argued to be telltale signs of endothermy. They involve the nostrils.

The nasal turbinates are thin wafers of bone or cartilage in the air passage of the nasal cavity that are lined with moist, mucus-producing epithelium. They act as heat exchangers by heating inhaled air and cooling exhaled air in birds and mammals. The mechanism of their function is explained in more detail in Appendix 2, but their principal function in modern mammals is to retain moisture from the warm, moist air coming from the lungs during expiration. The turbinates are so delicate that they do not fossilize, except under exceptional circumstances.

Glanosuchus was a therocephalian therapsid from the Late Permian of the Karoo—named by Robert Broom in 1904—that has been at the center of a long-standing debate about turbinates. Al Bennett was the first to suggest that, if nasal turbinates could be found in therapsids, this would be direct evidence that they were endothermic. Willem Hillenius from the University of California at Los Angeles argued that the skull of *Glanosuchus* showed signs of the remnants of nasal turbinates, which suggested a higher rate of respiration and oxygen consumption perhaps indicating the first real sign of endothermy in the mammal lineage.[6] Modern reptiles, such as lizards, do not have nasal turbinates, but birds do.

Animals that have short snouts, such as humans, do not have extensive nasal turbinates, yet they are highly endothermic. In this latter case, the same might be true of fossil animals that have traditionally been considered to lack turbinates. An excellent example involves dinosaurs. All early advocates of the turbinate idea—Al Bennett, John Ruben, and Willem Hillenius—maintained that dinosaurs were ectothermic and hence did not need to have nasal turbinates. But now there is evidence that not only were many dinosaurs endotherms but they did indeed also possess nasal turbinates.

Jason Bourke from Ohio University and six colleagues have shown that *Stegoceras validum*, an ornithischian dinosaur, possessed the bony ridge in the nasal cavity to which nasal turbinates were attached.[7] Given the lack of fossilized remains of the turbinates themselves, the existence of the remnant attachment ridges has been used as evidence of the existence of nasal turbinates in the therocephalian and cynodont therapsids.

The best recent evidence of nasal turbinates in the mammal lineage comes from the great survivor of the mass extinction at the end of the Permian, the dicynodont therapsid *Lystrosaurus*. This proto-mammal not only survived the extinction event but went on to become the dominant land animal in the awful environmental conditions of the Early Triassic. *Lystrosaurus* flourished—so much so that the first stratigraphic geological layer of the

Early Triassic in the Karoo Basin is known as the *Lystrosaurus* Assemblage Zone. *Lystrosaurus* flourished on the edge of the lethal tropics, at latitudes higher than 40° north and south, in the Karoo, Antarctica, and India.

Michael Laass from Martin Luther University and his collaborators employed neutron tomography to examine the snout of an exceptionally well-preserved fossil of a juvenile *Lystrosaurus declivis* from the Karoo.[8] Neutron tomography is similar to X-rays, but it penetrates rock better. The authors showed that the fossil had a complex network of cartilaginous turbinals that were well fossilized. These data are the best proxy yet for the telltale signs of an elevated metabolic rate in the mammal lineage. They provide firm evidence that endothermy had commenced at least 252 million years ago at the onset of the Triassic. A well-developed system of nasal turbinates would have allowed *Lystrosaurus* to minimize water loss during the hot and arid conditions of the Early Triassic.

Body size miniaturization occurred in the bird lineage as well, and much more dramatically. At the time of the mammal "exile" into the nocturnal world, a burst of endothermy occurred that was associated with extreme miniaturization and the doubling of relative brain size in nonavian theropod dinosaurs. Michael Lee from the South Australian Museum in Adelaide, along with collaborators from the United Kingdom, Italy, and Hungary, have shown that miniaturization occurred in the dinosaur lineage in the lead-up to the emergence of the birds and flapping flight. As they wrote: "There is a prolonged, directional trend in size reduction that spans at least 50 million years and encompasses the entire bird stem lineage from the very base of Theropoda, with rapid decreases in 12 consecutive branches from Tetanurae onward. . . . The ancestral tetanuran is inferred to be ~198 million years old and ~163 kg [359 lb], and size then decreases along subsequent nodes as follows: neotetanurans/avetheropods (~174 million years ago [Ma], ~46 kg [101 lb]), coelurosaurs (~173 Ma, ~27 kg [59 lb]), maniraptorans (~170 Ma, ~10 kg [22 lb]), paravians (~167.5 Ma, ~3 kg [6.6 lb]), and birds (~163 Ma, ~0.8 kg [1.8 lb])."[9]

By any standards, this was a profound decrease in the average size of a lineage of animals: from 163 kg (359 lb) to less than 1 kg (2.2 lb) over 35 million years, that is, nearly 5 kg (11 lb) per million years. Perhaps our fascination with dinosaur gigantism has muddied our thinking about underlying body size patterns that have occurred in dinosaurs.

However, a recent study on the relatedness between dinosaurs on the dinosaur family tree has rocked the paleontological world radically. It suggests that things were a bit different at "the very base of Theropoda" and that the Lee study tells only a part of the dinosaur miniaturization story. Contrary to the century-old traditional classification of Theropoda and Sauropoda as sister clades within Saurischia ("lizard hipped"), and Saurischia, in turn, as the sister clade to Ornithischia ("bird hipped"), in 2017 Matthew Baron and collaborators argued that Ornithischia and Theropoda are sister clades within a new group that they call the Ornithoscelida (arrangement shown in Figure A.4).[10] What makes this rearrangement highly noteworthy with respect to miniaturization is that all of the oldest theropods (Neotheropoda), such as *Eoraptor*, *Eodromaeus*, and *Tawa*, were small, omnivorous, bipedal, highly mobile dinosaurs that weighed less than 10 kg. Even the oldest ornithischian dinosaurs in the new placement, such as *Pisanosaurus* and *Eocursor*, were smaller than 10 kg. Moreover, the very oldest of all dinosaurs, the 247-million-year-old (Baron's date) ancestor of the sauropods, ornithischians, and theropods, was *Saltopus*, a tiny cat-sized dinosaur that weighed a mere 1 kg (2.2 lbs). *Saltopus* was also highly agile and bipedal and was less than 10 million years younger than the first archosauriform fossil exhibiting the first signs of bipedalism: *Euparkeria*. Many of these early theropods had grasping thumbs: they could use their hands to gather, catch, and hold food, and perhaps each other. So the real story seems to suggest that the body size of bipedal dinosaurs on the evolutionary pathway toward highly endothermic birds started small in the Early Triassic, got somewhat bigger toward the end of the Triassic, and again underwent a radical miniaturization during the Jurassic and the Cretaceous.

However, because miniaturization preceded the Late Triassic body size increases, I see no reason why McNab's Miniaturization Hypothesis cannot be applied rationally to the bird lineage as well. As the body size of the bipedal archosauriforms and dinosaurs got smaller during the Triassic, long before the advent of flapping flight, an increase in per-gram metabolism would have preserved the fitness benefits once enjoyed by larger-sized homeothermic archosauriforms.

But not all 10-kg bipedal theropod dinosaurs were endothermic. The Oviraptoridae dinosaurs (shown in Figure A.4) had a mass of about 10 kg (22 lb), and their radiation occurred about 170 million years ago, virtually at the time that miniaturization resumed for the second time in the bird lineage. The oviraptorids brooded their eggs. They showed extraordinary parental care by remaining on the nest to protect their eggs even during catastrophic sandstorms. Gregory Erickson from Florida State University and several collaborators analyzed the fossil remains of individuals of four oviraptorid species found in close association with their egg clutches.[11] They were interested in knowing whether the dinosaurs had reached adult body size by the time they started to breed. This is important information because it separates the reptiles, which display the ancestral life history characteristic of starting to breed before reaching adult size, from the modern birds, which show the derived characteristic of breeding only after adult size has been achieved.

The oviraptorids showed the reptile condition. Endothermy had yet to show the pulse toward the modern state — extreme miniaturization had yet to take place. These data compare nicely with the estimates of the body temperature of oviraptorids of about 30°C (86°F) obtained by Robert Eagle's research team, which show that the avian endothermic pulse had not yet occurred. The estimated body temperatures are intermediate between those of reptiles and birds, but showed that some degree of endothermy did exist because the body temperatures were higher than the estimates for the warmest phases in Mongolia, where they once existed. In other words, their body temperature was elevated above that of their surroundings but most certainly not by as much as that of modern birds.

As we did with the mammals, we also need to pin down the first appearance in the fossil record of some form of insulation that would have prevented the heat produced by the avian thigh muscles from being lost. If endothermic heat production did evolve as early as *Scleromochlus* and *Euparkeria*, we might expect that insulation should have evolved simultaneously or soon thereafter to minimize heat loss from the thighs and the rest of the body. Of course today birds have feathers of several kinds, some of which serve an insulating function, but I discuss this topic in more detail in the next chapter.

Let us finish this chapter by asking why miniaturization commenced so dramatically in the mammal and bird lineages during the Triassic.

In the case of the mammals, McNab argued originally that miniaturization occurred in response to competition with bigger therapsids, archosauriforms, and dinosaurs. However, the current consensus is that miniaturization was a passive process that occurred merely because the bigger-bodied clades of cynodonts went extinct during the Middle Triassic, leaving only small-bodied clades to persist into the Late Triassic. The same argument has been posed to explain miniaturization in the theropod dinosaurs,[9] so it seems that a common process may have occurred in both the sauropsid and the synapsid lineages but with different onset dates.

Adam Huttenlocker, then at the University of Washington, along with Jennifer Botha-Brink, studied the body size patterns of the Karoo therocephalian therapsids following the Permo-Triassic Mass Extinction.[12] The therocephalians were the most derived therapsids, but they went extinct in the Early Triassic. One therocephalian survivor of the mass extinction, the carnivorous *Moschorhinus*, showed a size reduction within the same species in the Early Triassic. In general, though, it was the small-bodied taxa that survived, for example, the therocephalian *Scaloposaurus* and the cynodont *Thrinaxodon*. The cynodonts radiated in size during the Early and Middle Triassic into bigger-bodied clades, but these clades went extinct, leaving only the small-bodied probainognathian cynodonts to inherit the Late Triassic and give rise to the Mammaliaformes and the mammals. So in the mammal lineage the onset of miniaturization can be pegged to at least the Permo-Triassic boundary 252 million years ago.

There is growing evidence that fast growth rates, early breeding, and early age of mortality were the adaptive life history characteristics in the highly unpredictable, semi-arid climate of the Early Triassic, at least as it is reflected in the rocks of the Karoo.[12-14] It takes time to grow big enough to breed, so Early Triassic therapsids would have faced high risks of dying before they could breed. The evolutionary solution to high rates of mortality is ubiquitous—breed early and die young.

Our fascination with enormous dinosaurs over the centuries has, unfortunately, clouded our understanding of why body size miniaturization occurred in the sauropsids. Unlike the synapsids, in which miniaturization was ubiquitous, miniaturized archosauriforms and dinosaurs coexisted alongside massive contemporaries—there was clearly no common selection pressure for body size miniaturization in the sauropsids. At the moment the best explanation is probably that small dinosaurs evolved to fill niches made available by related dinosaurs, such as sauropods, which were getting bigger and bigger.

The importance of miniaturization in the evolution of endothermy as we see it today in insulated birds and mammals cannot be understated. It was probably the single biggest driver of endothermy, and had it not happened, birds and mammals would probably not exist today.

SEVEN

Feathers and Fur

Miniaturization and endothermy are not good bedfellows. With miniaturization comes a vast increase in the ratio between surface area and volume, and hence larger relative areas of the body surface over which the precious new heat from the increased per-gram metabolic rate can be lost. Small, resting endotherms must be insulated if they are to maintain a constant elevated body temperature. However, those with ideas about the earliest stages of the evolution of endothermy will always be stymied by the lack of information about the integument of the therapsids and dinosaurs. Did therapsids have fat under their skins, like blubber? Did they have fur? Did the integument offer *any* form of insulation? Did dinosaurs have some form of insulation to trap the heat produced in their fat thighs? If so, what was it: fat, blubber, or perhaps some form of early proto-feather?

Danielle Dhouailly from the Institut Albert Bonniot in France proposed a model for the evolution of both hair and feathers.[1] She argued that the earliest amniotes had glands in their skin to keep it moist and slimy. A protein called alpha-keratin in the skin was laid down in layers in the epidermis and waterproofed the skin to some extent. When the amniotes gave rise to the synapsid and the sauropsid lineages, the glandular component was lost in the sauropsid reptile lineage. However, in the synapsid lineage, hairs evolved from the skin glands and, much later, so did mammary, sweat, sebaceous, and scent glands. Beta-keratin, on the other hand, evolved in the reptile lineage and was added to the alpha-keratin and lipids to make the animals' skins profoundly more waterproof.[2] Most glands were lost in the reptile integument. One was retained in birds, the single uropygial gland located on the "parson's nose" or tail, which secretes oils to keep feathers waterproof.

In modern mammals, insulation can be achieved in one of three ways: by blubber, fur, or both. These insulating structures trap pockets of air. Air is a very poor conductor of heat. Aquatic mammals such as whales rely on blubber, seals on fur and blubber, whereas land-based mammals are generally furry, except for the armored mammals and us humans—well, most of us. Some spiny mammals, such as porcupines and tenrecs, which have compromised insulation because their hairs have become quills, tend to have large subcutaneous fat deposits that compensate for the loss of insulation from thick fur. There is no known fossil evidence of fur in therapsids, but the possibility that they might have had fur, or perhaps a subcutaneous fat layer, can most certainly not be discounted.

Brian McNab has suggested that selection for fur in mammals probably commenced before the onset of miniaturization: "It is my conclusion," he wrote, "that the evolution of endothermy in the phylogeny of mammals occurred first by increasing thermal independence from the environment through an increase in mass, then by modifying the thermal properties of the surface, and finally by decreasing body mass with only a moderate reduction in the total rate of metabolism."[3] Let us examine this prediction.

The oldest unquestionable fossil evidence of fur occurred 165 million years ago, in the Middle Jurassic, in the swimming mammaliaform docodont *Castorocauda lutrasimilis* (Figure 7.1).[4] The figure presents an extraordinarily well-preserved fossil from China that shows the indelible outlines of a furry coat. A more recent 125-million-year-old Middle Jurassic docodont fossil of the arboreal 27-g (1-oz) *Agilodocodon* also exhibits evidence of fur.[5]

However, we can employ a trick in evolutionary biology known as homologous character state reconstruction to determine a more realistic date that is not pegged to a single fossil. The argument goes something like this: if *Castorocauda* was furry, it must have inherited furriness from the ancestral docodont. So, if we can pin a date to when the docodonts originated, we

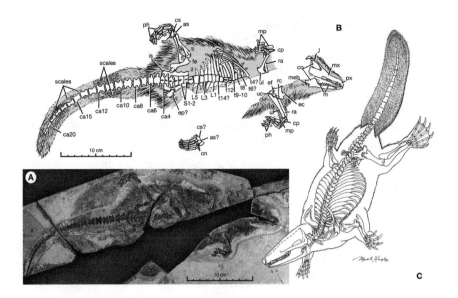

Figure 7.1. A photograph of the holotype—first specimen—of *Castorocauda* (A), a reconstruction of the bones and soft tissues (B), and a reconstruction of the animal as a swimming and burrowing mammaliaform (C). (Image courtesy of Zhe-Xi Luo, University of Chicago; published in Ji, Q., Luo, Z.-X., Yuan, C.-X., and Tabrum, A. R., "A Swimming Mammaliaform from the Middle Jurassic and Ecomorphological Diversification of Early Mammals," *Science* 311 [2006]: 1123–1127; American Association for the Advancement of Science License no. 3958120174497)

have the oldest estimated time for the first appearance of fur. It is a neat and simple way to estimate the time of the first appearance of fur.

Of course *Castorocauda* could, alternatively, have inherited fur from an immediate ancestor *within* the docodont clade. This would imply that the ancestor of the docodonts was not furry and that fur evolved twice independently in the mammals: once in the docodonts and again further down the line in their sister lineage. This alternative is unlikely but not impossible.

A conservative estimate of the time of divergence of the docodonts is about 230 million years ago, dated to the Late Triassic. The date can be interpreted based on a study by Zhe-Xi Luo from the University of Chicago and his collaborators, who described the lower jaw of another early mammaliaform, *Haramiyavia clemmenseni*, from East Greenland.[6] *Haramiyavia* itself was about 208–201 million years old, but the authors dated the divergence of the haramiyidans from the mammaliaform lineage closer to about 230 million years ago. I say that the age may be conservative, because these authors place the haramiyidans in their reconstructed mammaliaform phylogeny in a more derived position than that of the docodonts. In other words, the docodont origin was slightly older than that of the haramiyidans, so the age of the first appearance of fur may be even older than 230 million years, falling perhaps within the Middle Triassic.

There is an alternative approach to pinpointing the first appearance of hair, and it involves the analyses of genes. There are about four hundred genes that have been identified in mice that make hair happen—the "hair genes." Of course, if we had DNA from therapsids we'd be able to see whether these genes existed in these proto-mammals. But we don't, so other approaches are needed.

Craig Lowe from Stanford University and four collaborators have tried to pinpoint the timing of the emergence of hair by tracing the existence of the mouse hair genes through the animal kingdom.[7] They did this by looking for the sections of the DNA that are involved in regulating genes engaged in making hair proteins. These are called conserved nonexonic elements (CNEEs). CNEEs can produce new innovations, what geneticists call "novel phenotypes," by regulating the expression of existing genes. The trick is to look for regions of the DNA where there is enrichment of the CNEEs, and then look for the nearest functional genes.

The study by Lowe and others showed that the greatest enrichment of regulatory regions of the DNA occurred in the Triassic, at exactly the time that the cynodonts were radiating after the Permo-Triassic Mass Extinction. They also showed that weak enrichment commenced as early as the Carboniferous, which suggests that the earliest amniotes might have expressed some or other structure associated with hair. These early structures were probably skin glands, which secreted phospholipids onto the skin to render it more waterproof during the period of the earliest haul-outs of stem amphibians from the Carboniferous swamps. Hair evolved from these skin glands much later in the mammal lineage.

There is no reason why Brian McNab's Miniaturization Hypothesis cannot be applied to the theropod dinosaurs as well. Despite the wobbly pattern of size increases and decreases that were shown by the early theropods, big or small, they had the potential ability to generate heat from their big thigh muscles. The biggest would also have enjoyed a certain degree of homeothermy courtesy of thermal inertia. However, homeothermy and all its fitness benefits, whatever they may have been, would have been lost with miniaturization unless the per-gram metabolic rate was increased, for example, through increased leakiness of the mitochondrial membranes (described in Appendix 1). Again, as with the argument posed for the therapsids, we can then also expect some form of insulation to have evolved more or less in concert with miniaturization to prevent heat loss from the body.

After the pterodactyls split off from the reptile lineage, the remaining family tree consisted entirely of dinosaurs and their descendants, birds. The dinosaur lineage is stuffed with hard-to-remember clade names, so I'll tone down their use somewhat in hopes that they don't cloud our story as we move toward fully endothermic birds. As I explained in the previous chapter, current wisdom has it that Dinosauria split into two major groups: Ornithoscelida (Ornithischia and Theropoda) and Saurischia (the Sauropoda).[8] We are most interested in the Theropoda, the theropod dinosaurs, because it was they who gave rise to the modern endothermic birds, although this does not mean that features of endothermy did not appear as well in the Saurischia and the Ornithischia.

The Theropoda later split into the Tetanurae and the Ceratosauria. After the Tetanurae emerged during the Early Jurassic, 199 million years ago, they diversified very rapidly about 25 million years later into a multitude of smaller theropod types, including, ultimately, we think, the birds.

I must emphasize here that the above phylogenetic scenario indicating that birds evolved from theropod dinosaurs is by far the most widely accepted. However, Alan Feduccia has argued fiercely and relentlessly that the birds evolved not from the theropod lineage but rather from an archosauromorph somewhere before the emergence of Dinosauria.[9] Moreover, he argues that certain theropod clades most closely related to true birds, such as the well-feathered ostrich-like Ornithomimosauria,[10] are merely secondarily flightless birds and not true theropods. Feduccia also does not believe that feathers evolved for the initial function of providing insulation in theropods. Along with Theagarten Lingham-Soliar from Nelson Mandela University, Port Elizabeth,[11] Feduccia sustains the argument that all reports of early feather forms in non-maniraptoran dinosaurs are false: the fossil imprints merely represent collagen fibers produced during the degradation process rather than evidence of feathers.

Although these counteropinions should never be dismissed, for the purposes of continuity I will avoid getting too involved in the dispute and follow the prevailing wisdom. The debate does not affect my story.

Feathers were the obvious source of insulation in the dinosaurs because birds are, of course, feathered, and birds evolved from dinosaurs. So during the evolution of the dinosaurs, or perhaps even in their immediate archosauriform ancestors, somewhere, at some time, some proto-feather emerged from the reptilian/dinosaurian skin. Whether or not its initial function was to provide insulation is a highly fractious debate, as I mentioned earlier, because alternative functions such as display—colors and shapes—for mating rituals, warning signals, tactile sensing, or perhaps even camouflage, have been postulated.

Walter Persons and Phil Currie from the University of Alberta, Canada, have argued that feathers evolved initially from neurosensory bristles, such as can be seen today around the bills of many modern birds, and they dismiss a possible subsequent role in insulation.[12] However, more general explanations argue that feathers developed from the skin over all parts of the body and can be seen in various stages of development throughout the dinosaur lineage.

Feathers are quite a remarkable evolutionary innovation. They are made of beta-keratin and are both extremely lightweight and strong for their size. In a modern bird's feather, barbs branch out laterally along the length of a central shaft called the rachis. On the ends of the barbs are barbules that hook onto the barbules of adjacent barbs to form the feather vane. This is called a pennaceous feather. However, at the base of some feathers the barbs do not possess barbules, so they do not hook onto other barbs and are free. This gives the bases of some feathers their fluffy, downy look, and it is this part of the feather that is particularly good at trapping air and providing such excellent insulation. A downy feather on its own, that is, without the pennaceous bit on the top end, is called a plumulaceous feather.

Feathers develop from surface podlike condensations of beta-keratin cells on the skin called placodes. Richard Prum from the University of Kansas and Alan Brush from the University of Connecticut proposed distinct feather stages that progress from a simple, hollow shaft (stage 1) to the most advanced of all feathers, the asymmetrical, vaned, pennaceous feather (stage 5) that modern birds typically possess.[13,14] A feather bud or "germ" emerges from the placode and ultimately elongates into a tube. At the base of the feather germ develops the feather follicle, which possesses keratin cells that grow and force older cells up and out to form the shaft. In stage 2 the follicle cells differentiate into an inner row of cells that become the barbs and an outer layer of cells that make the protective sheath of the feather. Stage 2 looks like a feather duster of barbs because it does not have much of a rachis. Stage 2 is the downy feather, the heat-trapping feather. Stages 3 and 4 are vaned feathers, but they are symmetrical, which means that the vane on each side of the rachis is the same width. These feathers are called contour feathers because they cover the body of the animal. Stage 5 is the most advanced feather stage; it is vaned and asymmetrical—that is, one side is narrower than the other. These feathers typically occur on the wings and are critical for providing the aerodynamic properties needed for flapping flight.

Using the Prum and Brush model of feather development, we can trace the various developmental feather stages back through the theropod dinosaur lineage that led to the birds. We can do this because, remarkably, feathers fossilize despite their delicate nature.

Feather fossilization is made possible by the presence in the feathers of melanosomes, organelles that synthesize and store melanin, the most common light-absorbing pigment in animals. Melanin gives animals their coloration and protects the skin from harmful ultraviolet radiation. Although melanosomes are all cigar-shaped and roughly 500 nm (20 µin) long, their shapes vary relative to the colors that they absorb. Variation in melanosome shape throughout the feathered body of *Anchiornis huxleyi* has allowed paleontologists to reconstruct the colors of the plumage of this long-extinct bird. Placed within Troodontidae, the immediate sister clade to the true birds, *Anchiornis huxleyi* was the first feathered dinosaur to have its plumage color reconstructed based on an analysis of the feather melanosomes (Figure 7.2).[15] It is remarkable that it is possible to reconstruct the colors of a feathered dinosaur that inherited its color and feather characteristics from an ancestral troodontid that first appeared about 160 million years ago. *Anchiornis huxleyi* had a gray body, a rufous crest, and black wings with white stripes. It also had leg "trouser" feathers that were the same color as its wings.

Anchiornis huxleyi was one of the many exquisitely preserved Chinese fossils of feathered theropods and early birds that were surprisingly well articulated—that is, intact, showing that they were buried very quickly following their deaths, preventing their dismemberment by scavengers. They were also buried in shoreline silt deposits, so they were fully wetted immediately

Figure 7.2. The feather colors of *Anchiornis huxleyi*, named in honor of the influential biologist Thomas Huxley, were reconstructed from the size of the fossilized melanosomes in its feathers. (Image by and courtesy of Michael DiGiorgio, http://www.sciencecodex.com/ yale_scientists_complete_color_palette_of_a_dinosaur_for_the_first_time)

after death. The wetting itself would have had a definite influence on the shape and look of certain feathers, especially the downy body feathers. During their decomposition by bacteria, the beta-keratin in the feathers would have disappeared faster than the melanosomes, which are decomposed with more difficulty. So those sections of the feathers that did not have any color, such as the base of each feather, the calamus, which is embedded into the epidermis of the skin, would have decomposed first. Such differential decomposition of the parts of the feathers resulted in distorted appearances of the feathers in fossils because some remnants avoided complete decomposition and went on to become mineralized, whereas others did not.

What we can interpret about feather structure from the fossils can therefore be confusing. Over the vast period of time these animals spent lying in their sediments, they were compressed and mineralized into stone. They were squashed flat, very flat, which could have changed the shapes of feathers from what they looked like when the animals were alive. This problem concerned Christian Foth from the University of Rostock in Germany so much that he took a European siskin and squashed it in a printing press to see what the feathers looked like afterward, and concluded: "Flattening of specimens during fossilization amplifies the effect of overlapping among feathers and also causes a loss of morphological detail which can lead to misinterpretations."[16]

Indeed.

Another common outcome of the differential decomposition of beta-keratin and melanosomes is that a feather that might have had barbs and barbules ended up looking like a filament because all that fossilized was the rachis. Foth has reinterpreted several feather structures in fossil theropod dinosaurs and early birds that he claims were assessed incorrectly because of this problem.

The first feather-like structures, simple filaments probably equivalent to stage 1 feathers, evolved in the Early Jurassic and can be seen in a fossil of *Heterodontosaurus tucki* from the Karoo (Figure 7.3). *Heterodontosaurus* was an unusual dinosaur because it had differentiated teeth—hence its name—which most other dinosaurs do not have. It was tiny, as dinosaurs go, less than 1 m (3.3 ft) long, about the size of a smallish dog. It was a plant

Figure 7.3. Heterodonty (different teeth) is a rarity among dinosaurs, which generally have homodont (similar) teeth. *Heterodontosaurus*, as its name reflects, was a herbivorous ornithischian dinosaur that sported a body covering of filamentous spines. (Image by and courtesy of FunkMonk [Michael B. H.] CC BY-SA 3.0, https://en.wikipedia.org/wiki/Heterodontosaurus#/media/File:Heterodontosaurus_restoration.jpg)

eater, and in its hind leg the tibia was longer than the femur, which imme-diately suggests that it could run fast. It had filamentous feathers seen in the fossil as a crest along the dorsal surface of the tail but probably covered most of the body. So although *Heterodontosaurus* was an ornithischian dinosaur and not a theropod, we can probably peg the first proto-feathered dinosaur to an age of about 200 million years.

However, as I intimated earlier, the best fossil evidence of feathers in dinosaurs comes from China and not the Karoo. One site in particular, the Yixian Formation in Liaoning Province, China (from 124 to 128 mil-lion years ago, in the Early Cretaceous), has produced the most spectacular wealth of feathered dinosaurs in the world.[17] The first fossil from the site to show feathers, reported in 1997, was *Sinosauropteryx prima* (Figure 7.4).[18] It is placed in the family Compsognathidae within the Coelurosauria, the theropod clade that includes the true birds, Avialae. *Sinosauropteryx prima* weighed about 500 g (18 oz) and was about 1 m (3.3 ft) long. Most of its length, though, was its tail, which was longer relative to its body than that of any other theropod. It also had short front legs and long hind legs with enor-mous thigh muscles. The type of feathers that *Sinosauropteryx* possessed is debatable; they were definitely tubular, but it is not certain whether

Figure 7.4. Reconstruction of *Sinosauropteryx prima* (left) by Alexander Lovegrove. This fully feathered dinosaur included archaic mammals in its diet. The bands of feathers on the tail contained melanosomes that gave the feathers a brownish color; the feathers between the brown bands were probably white. The gliding lizard is *Xianglong zhaoi*. (Image by and courtesy of Alexander Lovegrove, http://alexanderlovegrove.deviantart.com)

they were downy or vaned. Interestingly, *Sinosauropteryx* fossils provide evidence that dinosaurs hunted and ate archaic mammals. One fossil contained mammalian jaws, those of a multituberculate, *Sinobaatar*, and a spalacotheriid, *Zhangheotherium*. Both of these mammals evolved after the monotremes but before the placental and marsupial mammals (shown in Figure A.3). They were small, nocturnal mammals. It was likely, therefore, that *Sinosauropteryx* was also nocturnal and hunted at night, perhaps due to being endothermic and having insulation provided by its feathers.

In 1998, one year after the discovery of *Sinosauropteryx*, two more feathered turkey-sized oviraptorosaurid feathered dinosaurs from the site were described: *Caudipteryx zoui* and *Protoarchaeopteryx robusta*.[19] These feathered dinosaurs were more closely related to true birds than was *Sinosauropteryx*, but they, too, could not fly. They did, however, possess both plumulaceous (stage 2) and pennaceous (stage 4) feathers. *Caudipteryx* means "tail feather" and was so named because of the long pennaceous feathers on its tail.

The discoveries from the Yixian Formation kept coming, and in 1999 yet another sensational feathered dinosaur was reported: *Beipiaosaurus inexpectus* (Therizinosauria), whose species name has immortalized it as the "unexpected" dinosaur. *Beipiaosaurus* was about the size of an ostrich and is one of the largest feathered dinosaurs known. We shall see why this feathered dinosaur was so unexpected shortly.

The Dromaeosauridae is a sister clade to the paired clade of Avialae and Troodontidae, and dromaeosaurid species such as *Microraptor* and *Sinornithosaurus* possessed both plumulaceous (stage 2) and pennaceous (stage 4) feathers that covered their whole bodies. *Sinornithosaurus* seems to have been unusual among feathered dinosaurs—it had fangs and is thought to have been venomous.

The feathered dinosaurs described so far from the Yixian Formation were all coelurosaurian dinosaurs, that is, they all occurred in the same clade as that of the true birds. However, this does not mean that phylogenetically older theropod clades were not bearers of some or other types of feather. *Sciurumimus albersdoerferi* was a noncoelurosaurian dinosaur that has been reconstructed with fluffy down (Figure 7.5). Its name means "squirrel mimic," because it had a long bushy tail covered in proto-feathers.[20] It is placed in the Megalosauria, one of the last nonavian dinosaur groups, which existed about 154 million years ago during the Late Jurassic before the emergence of the bird-theropods, the Avetheropoda. The fossil was found in stunningly well-preserved condition in a limestone quarry in Bavaria, Germany.

Perhaps the biggest surprise was the report of a small dinosaur with downy stage 2 feathers that was in the same clade as *Tyrannosaurus rex*. *Dilong paradoxus* (Figure 7.6) has been placed at the base of the Tyrannosauroidea family tree. It was about 2 m (6.6 ft) long fully grown, about the size of a largish dog. Of course, it could not fly.

What is notable about *Dilong* is that it was tiny compared with the enormous sizes attained in some of the tyrannosauroids, such as *Tyrannosaurus rex*, higher up the tyrannosauroid family tree. As the more derived tyrannosauroids got bigger, they lost their feathers; they were all quite defi-

Figure 7.5. Reconstruction of baby *Sciurumimus albersdoerferi*, a megalosaurid dinosaur, extracted from a limestone quarry in Bavaria, Germany. (Reconstruction based on Rauhut, O. W., Foth, C., Tischlinger, H., and Norell, M. A., "Exceptionally Preserved Juvenile Megalosauroid Theropod Dinosaur with Filamentous Integument from the Late Jurassic of Germany," *Proceedings of the National Academy of Sciences* 109 (2012): 11746–11751; image from http://upload.wikimedia.org/wikipedia/commons/c/c5/Sciurumimus.jpg, CC License)

nitely scaly.[21] I would argue that this example of what happened in the tyrannosauroid illustrates the close association between body size miniaturization and the concomitant need for insulation once any degree of internal heat production had evolved. And this is why *Beipiaosaurus inexpectus* is so unexpected—it was a largish dinosaur, and yet it was feathered.

The unexpected, though, can probably be explained by the climate of the Early Cretaceous around 145 million years ago; Eastern China at that time was actually quite cold and climatically variable. In collaboration with colleagues from the Chinese Academy of Sciences, Romain Amiot from the French National Centre for Scientific Research in Paris has used the same techniques that were used to estimate the temperatures of therapsids and

Figure 7.6. Reconstruction of *Dilong paradoxus*, a small tyrannosauroid, theropod dinosaur from the Early Cretaceous placed in
the same clade as the iconic *Tyrannosaurus rex*. The only fossil yet
discovered was probably a juvenile and displayed a mixture of pebbly
scales and filamentous proto-feathers. (Reconstruction based on
Xu, X., et al., "Basal Tyrannosauroids from China and Evidence for
Protofeathers in Tyrannosauroids," *Nature* 431 [2004]: 680–684, image
by and courtesy of Nobu Tamura, http://ntamura.deviantart.com)

archosauromorphs to reconstruct the Early Cretaceous climate of Eastern
China.[22] The method estimates environmental temperatures by measuring the stable oxygen isotope composition of apatite phosphate from the
remains of ectothermic reptiles. During the Early Cretaceous the Yixian
Formation was situated at about 42°N and the average temperature was
about 10°C (50°F). It was a temperate climate that had distinct wet and
dry seasons. So perhaps largish-feathered dinosaurs such as *Beipiaosaurus
inexpectus* would have had the ability to exploit higher latitudes seasonally,
for example, during the cold seasons, courtesy of their higher degrees of
thermal inertia and insulation.

The Tetanurae is the one dinosaur clade that includes all species that we
have discussed, for example, those from Megalosauridae, Tyrannosauroidea, Compsognathidae, Therizinosauria, Oviraptorosauria, Dromaeosauridae, and Troodontidae. This huge theropod clade of bipedal, carnivorous
dinosaurs first appeared about 200 million years ago. So if we take the divergence time of the Avialae as the time when feathered dinosaurs, that is,
true birds, started to use flapping flight—that is, around 160 million years
ago—we can reach a simple induction that feathers played some functional
role or another in small dinosaurs for at least 40 million years before they

were co-opted into an aerodynamic role for flapping flight. The term that evolutionists use for this is exaptation, the development of a feature that has a function for which it was not originally selected.

Christian Foth and his collaborators performed a character state reconstruction of the evolution of pennaceous feathers (stages 3–5), which led them to the conclusion that these feathers evolved in the ancestor of the Pennaraptora (Therizinosauria, Oviraptorosauria, Dromaeosauridae, Troodontidae, and Avialae).[23] This ancestor is dated to about 170 million years old.

Craig Lowe and his collaborators have also traced the "feather genes" through the animal kingdom. Again, as they did for the fur genes, they looked for the sections of the DNA that are involved in regulating genes concerned with making feather proteins, the CNEEs. In addition to finding that enrichment occurred adjacent to the feather genes, other surprises emerged: "Rates of innovation at feather regulatory elements exhibit an extended period of innovation with peaks in the ancestors of amniotes and archosaurs. We estimate that 86% of such regulatory elements and 100% of the non-keratin feather gene set were present prior to the origin of Dinosauria. On the branch leading to modern birds, we detect a strong signal of regulatory innovation near insulin-like growth factor binding protein (IGFBP) 2 and IGFBP5, which have roles in body size reduction, and may represent a genomic signature for the miniaturization of dinosaurian body size preceding the origin of flight."[24]

These authors argue that the regulatory section of the DNA adjacent to the feather genes was enhanced prior to the Dinosauria, that is, when *Eupakeria* was snatching "dragonflies" from the sky and skinny *Scleromochlus* was trying to be a pterosaur. So this date may not be dissimilar to that for the origin of hair in the mammal lineage. They also identify the accompanying genetic mechanism for body size miniaturization.

For me the argument about whether feathers evolved first to be used for functions such as display—in response to sexual selection—or to serve as camouflage, or perform some other nonthermoregulatory role, really amounts to premature squabbling. It is almost impossible to be certain without more data, much more data, about what the true situation(s) was or were. All modern birds rely on their feathers to stay warm, whereas not all of them rely on feathers to perform additional functions such as display, camouflage, or even, for that matter, flight. Stage 2 feathers preceded stage 5 feathers, so the insulation function of feathers arose very early during feather evolution in the Tetanurae. The average size of the Tetanurae decreased by a whopping 117 kg (258 lb) over a mere 25 million years from their first appearance 200 million years ago until the divergence of Neotetanurae about 175 million years ago.[25] Miniaturization was made possible by the accumulation of characteristics that enhanced endothermy: increased basal metabolism and heat-producing fat thighs, which allowed small dinosaurs to seek out new niches not occupied by their gigantic sauropod relatives.

The physical and genetic observations that this dramatic miniaturization occurred in concert with the evolution of insulating feathers is, I'd argue, excellent support for the idea that the principal focus of selection, at least for stage 2–4 feathers, was for insulation. I have never entered the feather function debate, and I would not, at this time, discount initial functions other than insulation. However, given the vehement argument in support of feathers' initial function having been for display purposes, I do wonder why large-bodied tyrannosauroid dinosaurs lost their downy feathers if this function apparently spawned important fitness benefits. When some dromaeosaurid feathered dinosaurs, which are more closely related to the true birds than were the tyrannosauroids, attained "large" sizes, they retained their pennaceous feathers, apparently for display and not aerodynamic purposes.[26] *Zhenyuanlong suni* was the size of a large turkey; it was about 1.7 m (5.4 ft) long and is estimated to have weighed 20 kg. But compared to *T. rex*, how big is "large"? In a 20-kg dinosaur living at 42°N in an average temperature of 10°C (50°F), I can well understand why feathers, whatever their stage, would have been retained for thermoregulatory purposes.

EIGHT

Taking to the Air

Pterodactyls were the first flying reptiles. It is hard to get one's head around the pterodactyls. The biggest, *Quetzalcoatlus*, had a wingspan of 12 m (39.4 ft), about the same width as that of a small airplane (Figure 8.1). No other animal bigger than *Quetzalcoatlus*, ever, has managed to get airborne. The dimensions of this reptile were truly staggering. It weighed possibly as much as 250 kg (551 lb). It could fly by flapping and gliding. It had a furry covering, the hallmark of a true insulated endotherm. Naturally, arguments about pterodactyl endothermy and flight abilities have been resolute.

Donald Henderson from the Royal Tyrrell Museum of Palaeontology in Alberta, Canada, has argued that *Quetzalcoatlus* was too heavy to fly and had become secondarily flightless, like an ostrich, ground-bound.[1] He estimated that it weighed more than 500 kg (1,102 lb). Henderson's claim, though, seems to be overshadowed by the estimates of other workers, especially those of Mark Witton from the University of Portsmouth, England, and Michael Habib from Chatham University, Pittsburgh. Witton and Habib claim that *Quetzalcoatlus* was an excellent flyer, using a combination of flapping and gliding.[2] They argue that *Quetzalcoatlus* would have used powerful flapping flight to get aloft, and then relied on thermals and gliding to gain altitude for long-distance travel.

If pterosaur experts are in agreement that strong muscles powered flapping flight, then I'm happy to argue that they *must* have been endotherms. Today there are no ectothermic vertebrates capable of flapping flight, and it is unlikely that there ever were any during the Mesozoic. Gliding yes, but

Figure 8.1. Named after the Mexican god of the air, *Quetzalcoatlus* was the largest pterosaur and the heaviest animal ever to have had the ability to fly. It was the size of an adult giraffe and lived in the Late Cretaceous. (Image by, courtesy of, and © Dmitry Bogdanov, http:// dibgd.deviantart.com)

not flapping. Lizards and even frogs can glide—they have gliding membranes—but they cannot flap their arms. Ectothermic vertebrates cannot employ sustained locomotor activity because they cannot provide oxygen to the exercising muscles fast enough. They can employ an anaerobic metabolism for bursts of activity, but this cannot be sustained, especially in the smallest ones, because of the rapid onset of blood acidosis. (The physiology of anaerobic metabolism during intense exercise is discussed in more detail in Chapter 16.)

It might be argued that the largest pterosaurs could have been flying ectotherms because all that was needed was a short burst of anaerobic flapping activity to get them aloft, and thereafter they could have restored blood acidosis during static gliding. Indeed, James Marden from Penn State University has argued this point, but he does not conclude that pterodactyls were necessarily ectothermic: "The problem of takeoff for the largest known flying dinosaur, *Quetzalcoatlus northropi* . . . is not vastly different than takeoff for a 1-kg [2.2-pound] vulture. Both should be able to take off via anaerobically powered flapping but need to reduce their power requirement, either by soaring or by flying at a forward speed where power requirements are below the level that can be sustained aerobically. In terms of ability to sustain flight, *Quetzalcoatlus northropi* does not appear to face problems any more severe than those for swans."[3]

Witton and Habib agree: "It is reasonable to expect that so long as giant pterosaurs launched within 1 to 2 kilometres [0.62–1.24 mi] of an external source of lift, they could then stay aloft by transitioning to a soaring-dominated mode of travel after an initial burst of anaerobic power."

Although endothermy in pterodactyls certainly does not preclude the use of anaerobic metabolism during takeoff, quite a strong case can be built against ectothermy as the condition of pterodactyls.

Pterodactyl bones were pneumatized, and their arms had a grossly extended fourth finger that supported a complex wing membrane. They had deep, sculptured sternums, allowing for the attachment of large pectoral muscles that powered the flapping downstroke. Indeed, Witton and Habib argued that the forearm possessed powerful adductor and abductor muscles.

From my own perspective, I'd argue that muscle temperature may have been important for takeoff performance. It is all very well arguing, in theory, that the largest pterodactyls could get aloft using an anaerobic metabolism only, but could pterodactyls of *all* sizes have done this with cool muscles? Modern lizards, except for the largest ones, such as the Komodo dragon, cannot employ burst speed when their muscles are cold, because they are small and do not have any capacity for thermal inertia—the capacity to retain heat. Crocodiles do, because they are big.[4] Cold flight muscles might not have grounded an animal the size of a 250-kg (551-lb) pterosaur because it would have been big enough to maintain a constant, probably elevated, body temperature through thermal inertia.

But can the same argument be applied to the smaller pterosaurs? Indeed, most pterodactyls weighed less than 1 kg (2.2 lb) and would have been easy prey for the myriad endothermic theropod nonavian dinosaurs if they were grounded during the cooler times of the day by cold flight muscles. Pterodactyls needed to be in a state of readiness for flight 24 hours per day, just as modern birds are today. The only way that the smallest pterodactyls could have achieved this was if they were endotherms and their flight muscles were perpetually warm and ready to generate power for takeoff and sustained flight.

Figure 8.2. *Sordes pilosus*, "filthy hairy," a small Late Jurassic pterodactyl with a wingspan of about 63 cm (24.8 in). The animal was
fully "furred" with pycnofibers. (Image by, courtesy of, and © Dmitry
Bogdanov, http://dibgd.deviantart.com)

The clincher in support of endothermy in pterodactyls is the evidence
that they were furry; they were covered in hairlike structures called pycnofibers. The best fossil evidence of a hairy pterodactyl is that of *Sordes pilosus*,
which means "filthy hairy" (Figure 8.2). It was a small pterodactyl with a
wingspan of less than 1 m (3.3 ft), unearthed in Kazakhstan.

True birds inherited the basic metabolic machinery needed to fuel the
huge expense of flapping flight. They also had warm flight muscles 24 hours
a day. Muscle-powered flapping flight was perfected during the Cretaceous.
It became possible because the bird *Bauplan* had acquired a threshold of
features that made flight possible: a three-fingered hand; a bipedal posture;
knee-based locomotion; pneumatic bones; vaned feathers; an increased
brain size, with an elaboration of the visual and olfaction regions of the
brain; a long, S-shaped neck; a wishbone (furcula); arm-folding ability; a
short, bony tail; extreme miniaturization; elongation and thickening of the
arms; and posterior orientation of the pubis and associated forward movement of the center of mass; as well as the partial fusion of the pelvic bones.[5]
All the right features needed for flapping flight were there.

Although miniaturization undoubtedly paved the way for flapping flight, pneumatic air-filled bones were also critical for ensuring not only a lightweight skeleton but also a reduced energy demand. Roger Benson from Cambridge University and his colleagues Richard Butler, Matthew Carranano, and Patrick O'Connor analyzed the occurrence of pneumatic (air-filled) bones in 131 nonavian theropod dinosaurs and found that postcranial pneumaticity first appeared in the Late Triassic, about 210 million years ago, long before the advent of flapping flight.[6] Apart from the not unexpected occurrence of pneumaticity in flying pterosaurs, it also occurred in the herbivorous sauropods, the largest animals ever to walk the earth. Roger Seymour has suggested that pneumatized bones in long-necked, giant sauropods would have increased their buoyancy and allowed them to float while feeding on the bottoms of water masses.[4]

With respect to the theropods, Benson and his colleagues have argued that the evolution of pneumatization "was likely driven by weight savings in response to gravitational constraints. Skeletal mass reduction in small, non-volant [flightless], maniraptorans likely formed part of a multi-system response to increased metabolic demands. Acquisition of extensive postcranial pneumaticity in small-bodied maniraptorans may indicate avian-like high-performance endothermy."[6] Remember that, in addition to their proposed aquatic behavior, sauropods still needed to walk on land to lay their cleidoic eggs.

Bone is metabolically active tissue, so in replacing parts of it with cavities of air, the total bone mass was reduced, which consequently reduced the metabolic demands on the animal. Energy could therefore be used for alternative functions, such as for endothermy—keeping the body warm and powering flight.

There has been considerable debate about how to define the first true bird. True birds in the sense of my trawl through the literature on the sauropsid family tree, especially the Theropoda, are those "feathered theropod dinosaurs" that were the first flyers. For decades the famous *Archaeopteryx* was accepted as the first flying bird, but during this time there were really no other contenders for the title.

During my visit to the Berlin Museum of Natural History I made a special effort to see one of the first fossils of *Archaeopteryx* that was discovered. It is housed in its own cubicle, big enough for about four people. I do not know whether the fossil is the real thing or a cast. It does not matter. You are allowed two minutes in the cubicle, with a few accomplices, with the cubicle doors closed, to savor possibly the most famous fossil in the world, original or not. What impressed me was its delicacy and how small it was. The nails on its toes are so fine you can barely discern them, and so, too, are its teeth. The whole skull is about as long as a shelled walnut. *Archaeopteryx* was a fragile dinosaur that could glide, but it was probably not capable of flapping flight.

Before I venture further, to the early birds, and since we've been talking about the activity patterns of dinosaurs, it is worth emphasizing here that once the feathered dinosaurs started to take to the air, they became exclusively diurnal, operating during the daytime only. Despite the fact that several pterosaurs were nocturnal, it would seem that most true early birds could accomplish accurate precision muscle-powered flying during the daylight hours only. Perhaps pterodactyls could accomplish night flying, because many foraged over water, mostly by gliding, and did not need to negotiate complex three-dimensional landscapes in the dark. Of course, modern birds such as owls and nightjars fly at night, but they evolved much later, during the Cenozoic, once the modern avian brain had increased considerably in size.

Much of the sparring in the bird-from-dinosaur debate revolves around whether *Archaeopteryx* is better placed in the clade Avialae ("bird wings"), which is defined as all theropod dinosaurs more closely related to modern birds than to the Deinonychosauria (a synonym of Dromaeosauridae), or whether it is better placed in Deinonychosauria, which would make it not a bird. Deinonychosauria is obviously the sister clade to the Avialae. The modern debate erupted when Xing Xu and his colleagues described a new *Archaeopteryx*-like theropod called *Xiaotingia zhengi* from China and placed it, along with *Archaeopteryx*, in Deinonychosauria, not Avialae.[7]

An immediate response from Michael Lee of the South Australian Museum in Adelaide and Trevor Worthy of the University of New South Wales in Sydney employed more robust maximum-likelihood and related Bayesian methods to reconstruct the origin of Avialae using the same data and showed that there was a higher likelihood that Archaeopteryx was an avialan.[8] They pointed out that, if Archaeopteryx was indeed a deinonychosaurian, flight evolved twice in theropod dinosaurs: once in the true ancestor of Avialae and once in Archaeopteryx within Deinonychosauria, whereafter it was lost in later deinonychosaurian species.

Unlikely.

Pascal Godefroit from the Royal Belgian Institute of Natural Sciences and his colleagues described Aurornis xui, a 160-million-year-old avialan that was about 10 million years older than Archaeopteryx, which they placed basal to it in Avialae.[9] They found weak support for Archaeopteryx as a deinonychosaurian. Aurornis, which means "dawn bird" and is from China, was about the size of a bantam chicken, about half a meter (1.6 ft) long. It had claws on its wings and a long, bony tail. So the Godefroit analysis at least placed Archaeopteryx back within the avialans, somewhere near the root of the tree.

However, Steven Brusatte from the University of Edinburgh, Scotland, and his collaborators subsequently turfed Aurornis and Xiaotingia out of Avialae and placed them in a new sister clade called Troodontidae.[5] So once again, it seems, Archaeopteryx is named the first bird. Whatever the case, though, in my family tree I have placed Aurornis and Xiaotingia within Avialae (shown in Figure A.4).

All archaic birds at the base of Avialae had teeth in their jaws that the birds in Aves, the clade in which the crown birds reside, have lost. Robert Meredith from Montclair State University, New Jersey, and his colleagues estimated that teeth were lost in the ancestor of the modern birds about 130 million years ago.[10] Modern birds use their gizzards to process food, as an alternative to chewing.

So how did bird flight evolve?

Once the pennaceous feather evolved, nonavian feathered dinosaurs possessed planar surfaces on their wings, legs, and tail that could certainly be used for early aerodynamic functions. However, with the exception of

Microraptor, nonavian theropods did not possess asymmetrical pennaceous feathers, suggesting that sustained flapping flight was not possible. There were also limitations in arm structure that prevented the wings from being lifted above the shoulders, so making an effective wing upstroke was not possible. Birds cannot gain height without an upstroke of the wing. So what might have been the earlier alternative "flight" options?

Several models have been proposed for the evolution of avian flight. The earliest was the Cursorial Model proposed by Samuel Williston in 1878. That model proposed that birds such as *Archaeopteryx* employed a series of jumps during fast running and used their extended wings for thrust. Although the basic model has been modified over the years, the main criticism of the model is that *Archaeopteryx* was too heavy to have managed to get aloft from the ground. It still had heavy teeth and a long tail, unlike the lightweight keratin beak and the short parson's nose of modern birds.

The Wing-Assisted Incline Running Model argues that birds used their flapping wings to help them ascend steep inclines and even vertical surfaces, much as modern ground-dwelling birds do (Galliformes, such as chickens and partridges).[11] In modern game birds the same behavior is used for general locomotion, predator escape, and reaching lofty roosting perches. Although the main criticism is roughly the same as for the Cursorial Model, namely, that the birds were too heavy for flapping flight, the behavior can be applied to the very lightweight chicks of ground-dwelling birds. So this model cannot be discounted if wing-assisted incline running gave the chicks of bipedal carnivorous feathered dinosaurs a survival edge.

The model that seems to be most favored, and the one I'd defend if forced into the debate, is the Arboreal Model, or the Tree Down Model, proposed by Othniel Marsh, the great American paleontologist from Yale, and one of the combatants in the famous "Bone Wars." Edward Drinker Cope, another great American nineteenth-century paleontologist, the other combatant, waged a long and acrimonious war with Marsh over the discovery and naming of dinosaurs. The Arboreal Model essentially argues that dinosaurs with feathers used their feathered wings, legs, and tails to glide from one tree to another, much as modern flying lemurs (Dermoptera) and flying squirrels do. During foraging, they would have used the sharp claws on their proto-wings and hind legs to climb tree trunks, and then glide to the lower levels of adjacent trees. I like this model simply because it explains why hand claws were retained in all feathered dinosaurs right up until the time when true flapping flight occurred. They still existed, for example, in *Archaeopteryx*.

Which bird, then, was the first true flapping flier? *Archaeopteryx*, the "first bird," had characteristics indicative of the aerodynamic ability to create lift, such as asymmetrical wing feathers. It also had broad wings that would have given it a low stall speed. Its trouser legs would have added to the lift, and might have been used as air brakes to slow it down during landing, as raptors do today. It lived in a forested tropical world, so these characteristics would certainly have allowed it to glide from one tree to another. But could it use flapping flight to get above the forest canopy?

In a letter to *Nature*, Amy Balanoff from the American Museum of Natural History in New York and her colleagues reported an analysis of the brain sizes of coelurosaurian dinosaurs.[12] They showed that the brain size of *Archaeopteryx* was more similar to that of other troodontids and dromaeosaurids than to true birds within Aves. They argue that if brain size might be a proxy for flight capacity and if members of *Archaeopteryx* could fly, so too could their closest dinosaur relatives. But the brain size of *Archaeopteryx* might suggest that it had yet to attain the necessary capacity for true flight, such as spatial perception and balance. *Archaeopteryx* also lacked the keeled sternum of modern birds, which is required for the attachment of the powerful pectoral flight muscles. It also retained the long dinosaurian tail and thus lacked a site for the attachment of muscles needed to control tail feathers. In modern birds, the dinosaurian tail evolved into the short, squat pygostyle (parson's nose), which a fan of feathers under exquisite muscular control are attached to and employed for maneuverability. The transition from the dinosaurian tail to the pygostyle involved the shrinking and fusing of the last few tail vertebrae into a single bone.

The verdict is still out on whether the first bird was actually capable of flapping flight. From most accounts it seems that, at best, *Archaeopteryx* was a weak flyer and, at worst, a good glider. The first true flyers were probably found in the earliest members of the Ornithoraces, Avialae, which possessed a keeled sternum indicative of attachment sites for the large pectoral muscles that power sustained flapping flight. Apart from the enantiornithine birds, which were thought to have been able but slow flyers, the earliest flying bird was probably something like *Ichthyornis*, a Late Cretaceous seabird (shown in Figure A.5). The story is often clouded, though, by the fact that birds reverted back to flightlessness, as did *Hesperornis*, which evolved soon after *Ichthyornis* and was a diving seabird.

Today birds can achieve astonishing feats of athleticism courtesy of feathers and their endothermy. In terms of the sheer distance they can cover, the yearly migrations undertaken by some nonpasserines are legendary and have recently been described in some detail. The Arctic tern, *Sterna paradisaea*, is the champion of all. This medium-sized bird weighs about 120 g (4.2 oz) and migrates yearly from its breeding grounds in Greenland to its austral summer feeding grounds in the Southern Ocean off the coast of Antarctica.

Carsten Egevang from the Greenland Institute of Natural Resources, together with international collaborators, managed to track individuals by fitting them with miniature "geolocators" that weigh a mere 1.4 g (.05 oz).[13] The data recorded are quite stunning. The average round-trip distance covered in a year is 70,900 km (44,055 mi)—the maximum recorded for a single bird was 81,000 km (50,331 mi)—made up of 34,600 km (21,499 mi) southbound, 10,900 km (6,773 mi) foraging for food in the Southern Ocean, and a return journey of 25,700 km (15,969 mi). On the southbound migration, the birds chose either the west coast of Africa or the east coast of South America to follow. However, when they headed back to their breeding grounds, which they did in more of a hurry, they took a single route up the center of the Atlantic Ocean. No other vertebrate on earth can travel such huge distances in one year.

Although migration was undoubtedly not the original driving force that led birds to take to the air, its modern employment has contributed to the massive diversity of birds, particularly of passerine songbirds—about 5,000 species—that exists today. By escaping the harsh, unproductive winters of the northern hemisphere, when it is hard to find food to meet the massive costs of endothermy, migratory birds can swap seasons within weeks by heading for the southern hemisphere.

Migratory birds tend to breed in the northern hemisphere in preference to the southern hemisphere for several reasons. First, there is more land in the northern hemisphere, which means more available habitat. Second, the climate in the northern hemisphere's summers is more predictable, so birds have a high probability of successfully raising one or multiple clutches. For example, El Niño has a much weaker influence on summer rainfall patterns in the northern hemisphere than in the southern. Third, the northern hemisphere is infinitely more fertile than the southern hemisphere, courtesy of hundreds of thousands of years of composting through glacial action during the ice ages. Just try to imagine what a block of ice 1 km (0.63 mi) thick would have done to a boreal forest.

Endothermy underlies the success of birds and mammals today, certainly in terms of diversity and geographical distribution. Endothermy is indeed expensive, but natural selection has found ways for endothermy to make fitness profits relative to the costs during the dynamic climate changes that characterized the Cretaceous and especially the Cenozoic. Endothermy enabled muscle-powered precision flapping flight in birds. I find it hard to live without birds in my life. The garden behind my house is like a giant open aviary, a daily window into avian hot-bloodedness. My favorite is the Cape robin-chat, *Cossypha caffra*, which wakes me up every morning with its rambling song, and it always starts on the same note—stoic reliability—endothermy at its best.

NINE

The Heat of Darkness

The traditional account of the fate of the Late Triassic and Jurassic mammals has it that they were hustled into darkness by the riotous profusion
and ecological dominance of the dinosaurs. Whereas many paleontologists
would argue that this perception is not quite correct, what is agreed upon
is that mammals became true thermoregulating, endothermic mammals in
the dark, an idea first proposed by Alfred Crompton and his colleagues from
Harvard University.[1] The evidence is overwhelming—Mesozoic mammals
were creatures of the dark. The best evidence comes from studies on the
structure and function of the eye—how a day eye turned into a night eye—
the condition that still exists in mammals to this day, with some exceptions,
such as in anthropoid primates.

Derived cynodonts, mammaliaforms, and Mesozoic mammals were nocturnal. Why they became nocturnal during the Mesozoic is not known;
there has never been a succinct hypothesis on the topic. Suggestions are
limited to inferential paleontological arguments concerning niche occu-
pancy by diurnal dinosaurs. For example, Crompton and his colleagues argued that, given the capacity to defend a constant body temperature against
cooler nighttime air temperatures, ancestral mammals exploited an insectivorous, nocturnal lifestyle that had yet to be exploited by reptiles in the
sauropsid lineage.[1]

The evidence that nocturnalism was indeed the prevailing condition by
the onset of the Jurassic is growing. Excellent empirical evidence comes

from comparative studies on the relative sizes of brains. The increase in brain size was associated with increases in the size of the olfactory cortex and olfactory bulb, neocortex and cerebellum. The neocortex stores maps and patterns of vision, touch, hearing, and smell—critical innovations for a nocturnal existence—which are representations of the mammal's inter-action with its environment.[2] Once the full complement of jawbones had completed the evolutionary migration to the middle ear—to assume their roles as ossicles—those parts of the neocortex associated with the process-ing of sound patterns were enhanced. The ossicles are some of the smallest bones in the body and transmit sound from the air to the cochlea. The neo-cortex is thought to have evolved in tandem with insulated endothermy "to assure the reliable and continuous location of food resources."[2]

Nonanthropoid mammals have "nocturnal eyes."[3] The "nocturnal bot-tleneck" hypothesis argues that the switch by mammals from a diurnal to a nocturnal existence resulted in several changes to the ancestral optic condi-tion. Photic input pathways to the circadian pacemaker system in the brain were restricted to retinal inputs. In reptiles, dinosaurs, and birds there are photo input systems from the pineal gland and the parietal eye in addition to parallel inputs from the retina. The parietal eye, or "third eye," sits in a hole in the roof of the skull and allows light to penetrate directly through to the pineal gland in basal synapsids and sauropsids. It provides photic input to the pineal gland for body temperature control and seasonal and reproductive synchrony. Mammals have lost the third eye primarily because their large, highly folded neocortex smothered the pineal gland, but also be-cause, once nocturnal, they no longer needed to thermoregulate by moni-toring the time of the day to maintain synchronicity with the heat cycles of the day. Reliance upon this extraretinal photic input became redundant in nocturnal mammals following the evolution of endothermy.

Because the parietal eye left traces of its existence in the fossil record, its last appearance can be traced to the cynodonts, let's say to the Middle Triassic.[4] The Probainognathia, the lineage of the most derived Late Trias-sic cynodonts, from which the mammals evolved, had lost the parietal eye, suggesting a timeline for the switch to nocturnalism—the Late Triassic.

Nocturnal mammals display a greatly reduced number of cones relative to rods in the retina, indicating visual acuity in favor of light sensitivity; cones detect color, whereas rods work better under conditions of low light. In addition, two of the four opsin photopigments within the cones that are necessary for full color vision were lost.[5] The majority of living nonprimate mammals therefore are dichromatic and have limited color vision. A third cone photoreceptor was regained through gene duplication in primates— old-world monkeys and apes—and also some marsupials; this cone restored trichromatic, color vision in the Cenozoic.

About half of the retinal opsins that are not involved in vision were also lost. Some of these opsins are involved in entraining the day-night cycle— the length of the days and the nights—to synchronize seasonal breeding, which was essential for amphibians, reptiles, and birds. Nocturnal mammals also lost the melanopsin-encoding *Opn4x* gene and hence, unlike reptiles, dinosaurs, and birds, cannot express photopigments within brain tissue or other nonretinal structures.[5]

Diurnal animals need to protect their eyes, especially the retina, from damage caused by low-wavelength ultraviolet light. Photolyase DNA enzymes make repairs, but these have also been lost in nocturnal mammals. The loss is taken as evidence of their restricted exposure to solar radiation.

Oil droplets in non-mammalian retinal photopigment cells, some of which are colored, act as wavelength filters and add to color discrimination. Color in the oil droplets was lost in the ancestral mammal, and the entire oil droplet was lost later in the marsupials and placental mammals. Oil droplets reduce the amount of light that reaches the photopigments, and it is for this reason that it has been suggested that they were lost in Mesozoic marsupials and placental mammals, namely, to increase the amount of light reaching the photopigments at night.

The shape of the eye and the orientation of the eye orbit in early mammals also became adapted to low light.[3] The mammalian eye orbit faces forward, allowing an overlap of the field of view of each eye and hence binocular vision. Increased binocular vision enhances light sensitivity, depth perception, and contrast discrimination. In the latter respect, for example, binocular vision allows camouflage patterns to be broken in low light; it is thus well developed in mammalian nocturnal predators.

Mammals had become exclusively nocturnal by about 230 million years ago.

The word "mammal" is derived from the Latin word *mamma* ("teat"). Mammary glands are one of the main characteristics of mammals but not the one that defines them as a single taxonomic entity, Mammalia. The defining feature, as I have emphasized in earlier chapters, is the presence of the full set of ear bones, the ossicles, in the middle ear, which give true mammals their extraordinary hearing ability, especially at high frequencies.

The mammary glands, of course, produce milk. There is an evolutionary association between the appearance of hair and the very earliest production of milk, known as proto-lactation. Lactation is an ancient condition of the synapsids. The modern mammary gland is thought to have evolved from skin glands associated with hair follicles.[6] The skin secretions in early synapsids probably involved fluids that prevented desiccation of parchment-shelled eggs and phospholipids to minimize water loss from the skin. Milk attained its modern composition during the Late Triassic. In all likelihood, producing the energy-rich milk would not only have required endothermy but also have enhanced it. Mammals got smaller during the Great Shrinking, and consequently, so did their eggs.[6] Small eggs produce small babies, so the critical importance of high-energy milk at this time was that it allowed the neonates of small cynodonts, mammaliaforms, and early mammals to grow quickly.

The egg-laying mammals, the echidnas and platypuses of Australia, are considered true mammals, yet they do not possess teats. They do, however, produce milk from mammary patches on the skin, and it is from these patches that milk is lapped up by baby monotremes. This is probably the way that baby cynodonts also fed. Marsupials and monotremes have a curved bone attached to the pubis called the epipubic bone. It is also found in mammaliaforms and derived cynodonts but not in placental mammals. The epipubic bones are thought to assist the mothers in supporting the mass of eggs and babies in a pouch, but these elements became redundant in placental mammals with the evolution of the placenta.

Before I examine the huge challenges faced by the mammals in the dark, let's recall and examine the characteristics of the most advanced cynodont whose descendants entered the dark. Let's call this derived cynodont "dog-tooth," which is what the name cynodont means. Let's consider it as a ge-neric ancestor of the true mammals. I have created this generic cynodont because what follows is necessarily speculative—there are no published ac-counts of how these archaic mammals may have lived. Permit me.

Dogtooth was small in size. The majority of species weighed less than half a kilogram (1.1 lb), and many weighed no more than 100 g (0.22 lb). Dogtooth ate insects and laid soft-shelled eggs. It probably cared for its ba-bies until they were weaned. It could produce milk from mammary patches that its babies would have lapped up from the folds of skin on its tummy. It is almost certain that dogtooth had fur, at least a covering of proto-hair, which served both an insulation and a waterproofing function. It could probably generate some endogenous heat at certain times of the year, for example, during the incubation period, but it was highly unlikely to have maintained a constant regulated body temperature throughout the year. Dogtooth had a brain less than a quarter of the size of those of modern mammals. It could not hear high-frequency sounds very well because the bones of its lower jaw had yet to complete their migration to the middle ear. Compared with mod-ern mammals, it also had a relatively poor sense of smell, poor eyesight, and an ancestral means of tactile sensitivity. Its motor coordination was capable, but clumsy.

Importantly, though, dogtooth could burrow and climb. It had the fin-gers and the toes to do so. Ultimately, I hypothesize that the retention of the digits was responsible for the survival of the mammals at the bound-ary between the Cretaceous and the Cenozoic 66 million years ago. I say this because the flatfoot condition that retained fingers and toes kept the mammals small prior to the extinction of the dinosaurs, and it also allowed them to climb, swim, and burrow, activities essential for surviving the first few hours of the infernos following the meteorite impact at the boundary. Remember, once the bird lineage deviated from the flatfoot, probably start-ing with *Scleromochlus*, it opened the gates for the evolution of large body sizes,[7] which ultimately, I suspect, was the underlying cause of the extinc-tion of the dinosaurs. (This fascinating topic of differential survival at the boundary is discussed in more detail in Chapter 10.)

Imagine what it must have been like for dogtooth to conduct its daily activities in total darkness. How did it find food? How did it find mates, its brothers and sisters, its parents? How did it find its way back home? Let's imagine the problems that dogtooth would have experienced after becoming nocturnal.

During the day it would have had no reason to maintain a warm body temperature, except during the breeding season. It could hide in small places, such as in burrows underground, in caves and rocky hideaways, under the roots of trees, in the leaf litter, or in a hole in a tree. It would have needed to maintain a minimal metabolic rate, enough to stay alive, and its body temperature would have fluctuated with that of its surroundings. In other words, it would have resorted to its ancestral ectothermic state even though it had the capacity to elevate its basal metabolic rate. But, as evening approached, it would have needed to kick-start its metabolism to raise its body temperature above the rapidly cooling late afternoon air temperature. Alternatively, if it became active at sunset and remained so throughout the night, it might even have been able to produce sufficient heat from the exercise of activity to keep its body temperature above that of the cooler evening air without having to produce any extra endogenous heat. If this had happened, its body temperature would have been pretty variable as its activity level ebbed and flowed.

During the breeding season, however, much larger energy demands would have been placed on dogtooth. An elevated, constant body temperature would have increased the rate of egg development and considerably shortened the incubation time of eggs.

Of course, we do not know how the egg-laying dogtooth behaved when breeding, but we can make some guesses based on modern mammals that still lay eggs, the echidnas and platypuses.

In the platypus, the egg develops for twenty-eight days inside the mother's reproductive tract and, once laid, is incubated by the mother for only about ten days. The delayed laying of the eggs in the platypus is different from what happens in birds that retain the egg internally for only one day and then incubate it for three weeks. In echidnas the laid egg is incubated in a pouch in which the hatchling also develops.

Gordon Grigg from the University of Queensland has studied echidnas throughout his long career and has proposed some highly insightful ideas about the evolution of endothermy in what he called "protoendotherms." Grigg has never defined precisely what he means by a protoendotherm, but I guess that it is any animal that displays the first characteristics of endothermy, such as therapsids, with their nasal turbinates. With his collaborators Lyn Beard and Michael Augee, Grigg has published excellent data on body temperatures recorded in echidnas housed in outside pens. These data show that the tightest, most constantly regulated, elevated body temperature in female echidnas occurred during the incubation period only. Indeed, these authors concluded, based on these data:

> that the evolution of endothermy is likely to have occurred via stages in which selective pressures favored the enhancement of elements present within reptiles and that the early "protoendotherms" were facultative rather than obligate endotherms, like the only known extant reptilian endotherms. . . . Echidnas have the advantages of endothermy, including the capacity for homeothermic endothermy during incubation, but are very relaxed in their thermoregulatory precision and minimize energetic costs by using ectothermy facultatively when entering short- or long-term torpor. They also have a substantial layer of internal dorsal insulation. We favor theories about the evolution of endothermy that invoke direct selection for the benefits conferred by warmth, such as expanding daily activity into the night, higher capacities for sustained activity, higher digestion rates, climatic range expansion, and, not unrelated, control over incubation temperature and the benefits for parental care.[8]

I agree entirely, but some clarification is needed here: the "reptiles" referred to are the cynodonts, dogtooth and its kin. Mammals are not reptiles.

The "enhancement of elements" statement is an important distinction, because it is the first suggestion that selection for a warm body in the ancestral mammals would have occurred in a *suite* of activities wherever warmth promoted the animal's fitness. Tom Kemp from Oxford University has also argued that no single cause can account for the evolution of endothermy. In his Correlated Progression Model of endothermy he argued: "Each structure and function associated with endothermy evolved a small increment at a time, in loose linkage with all the others evolving similarly. The result is that the sequence of organisms maintained functional integration throughout, and no one of the functions of endothermy was ever paramount over the others."[9] Yet the hypotheses for the evolution of endothermy that have been proposed at this time are all single-cause hypotheses.

The single-cause versus multiple-cause ideas about how endothermy evolved is one debate. Another is whether endothermy, once it had evolved, was adhered to for the next, say, 200 million years. In other words, this idea dismisses the possibility that endotherms could revert back to an ectothermic state at any time of the year. The alternative to this idea is that, in certain mammals, the ancient ectothermic characteristics were retained and employed in concert with endothermy to maximize fitness.

I take the latter stance, one of the essential theses of this book. I emphasize the idea that early mammals that had an endothermic capacity had no reason to employ that capacity during the daytime to maintain a constant body temperature if there was no fitness benefit to being warm, such as when they were not breeding. In other words, Grigg's protoendotherm was a creature that retained its ancestral ectothermic condition but employed it during the nonbreeding season only. It used its newly evolved capacity for endothermy facultatively during the breeding season for activities in which warmth promoted fitness. I like this idea immensely, because there are living mammals that do this exactly today. They provided the incentive for me to write this book, but I will tell their story later.

Dogtooth gave rise to a group of non-mammals, the mammaliaforms, which were the most derived of the non-mammal synapsids. They were similar to modern mammals in most respects except that the articular bone in the lower jaw had yet to give way completely to the dentary. The articular still hinged with the quadrate in the skull, so these two bones had yet to make their way into the middle ear to become the malleus and incus ossicles.

One of the most ancestral mammaliaforms was *Morganucodon* (Figure 9.1), named after Glamorgan in Wales, where many of the first fossils were found. *Morganucodon* was much like a modern shrew. It was small, about

Figure 9.1. A charming reconstruction of a pair of *Morganucodon*, a mammaliaform from the Late Triassic. This animal showed one of the first measurable brain-size pulses above that of the therapsids. (Image by, used with the kind permission of, and © Cristóbal Aparicio Barragán, http://www.seirim.net)

20 g (0.78 oz), nocturnal, and insectivorous, and it lived in burrows. It laid leathery eggs similar to those of modern platypuses and echidnas. But it displayed an innovation that marked a huge new leap toward mammalness: it had an enlarged brain.

Our understanding of *Morganucodon* and her descendants has improved exponentially over the past 20 years due to excellently preserved mammaliaform fossils that have been found in northwestern China.[10] These fossils have been used to reconstruct some of the features that help to resolve the timeline for the evolution of endothermy. Undoubtedly the most revealing feature has been the leap in brain size.

Timothy Rowe from the University of Texas at Austin and his colleagues have used high-resolution X-ray computed tomography to scan the fossil skulls of various cynodonts; a few mammaliaforms, including most notably *Morganucodon*; as well as living and extinct mammals.[11] Their unit of comparison is called the encephalization quotient, which is a measure of the sizes of the brains after correcting for the different body sizes of the animals.

This technique allows the brain sizes of the smallest animals to be compared with those of animals of bigger body sizes. A typical modern mammal would have an encephalization quotient of about one. The data from the Rowe study led to conclusions not too flattering toward the cynodonts: "Compared with their living descendants, early cynodonts possessed low-resolution olfaction, poor vision, insensitive hearing, coarse tactile sensitivity, and unrefined motor coordination. Sensory-motor integration commanded little cerebral territory."[11]

Cynodonts had encephalization quotients of around 0.16–0.23, that is, brains less than a quarter the size of similar-sized modern mammals. Dogtooth would have been pegged at the upper limit of this range. Comparisons of the sizes and structures of the brains of cynodonts with those of mammaliaforms and true mammals allowed the Rowe team to pinpoint exactly *which* regions of the brain became enlarged. Brain expansion occurred primarily in the olfactory bulb, neocortex, olfactory cortex, and cerebellum. The pulses occurred in three waves.

Evidence of the first wave can be seen in *Morganucodon* in a pulse in the encephalization quotient from the upper limit of 0.23 in the advanced cynodonts to 0.32, an increase of nearly 50 percent. The areas of the brain that increased in size were associated with increased resolution in olfaction (smell), tactile sensitivity from body hair, and neuromuscular coordination.

The next encephalization pulse is evident in an extraordinary Jurassic fossil discovered in the Lower Lufeng Formation of Yunnan in China, *Hadrocodium*, a mere 2-g (0.07-oz) mammaliaform about the size of a paper clip and 195 million years old.[12] Its body size was equal to the lower limit that any endotherm, both living and extinct, has ever attained. It is as small as a mammal or bird can get (Figure 9.2). *Hadrocodium's* brain size entered the lower limits of the range that we see in modern mammals, with an encephalization quotient of 0.50. The new increase in the brain size occurred in the expanded olfactory bulbs and olfactory cortex. *Hadrocodium* must have had a vastly increased capacity to smell its environment.

The appearance of the neocortex in the mammaliaforms is a truly significant shift toward mammalness. Among the vertebrates, it is only the mammals that have a neocortex, the biggest part of the brain, which sits on top of

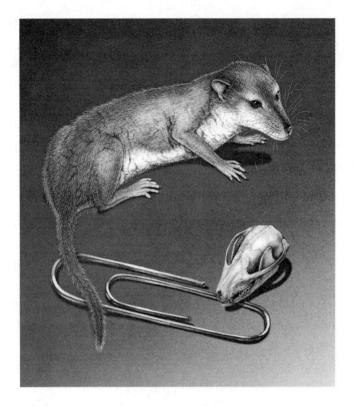

Figure 9.2. Reconstruction of *Hadrocodium* by Mark Klinger, Carnegie Museum of Natural History. This tiny mammaliaform had a
brain size that fell within the range of those for modern mammals.
(Image from http://www.aaas.org/news/science-sniff-sniff-smelling-led
-smarter-mammals; based on the study by Rowe, T. B., Macrini, T.
E., and Luo, Z. X., "Fossil Evidence on Origin of the Mammalian
Brain," *Science* 332 [2011]: 958–960; printed with the permission of the
American Association for the Advancement of Science [License no.
395812118213o])

the ancient reptilian brain. It is composed of gray matter—neurons that do
not possess myelinated fibers. It is the most derived part of the brain. John
Allman from the California Institute of Technology has argued that the
development of the neocortex must have occurred in tandem with endothermy "to assure the reliable and continuous localization of food resources.
The neocortex consists largely of topographically organized maps of vision,
touch and hearing that store representations of the mammal's interactions

with its environment. The neocortical network requires an initial period of training, which is manifested in infant play behavior. During this training period the infant's energy requirements are provided by its mother's milk. Thus mammalian lactation and play behavior are necessary adaptations to support linked endothermic and neocortical homeostatic mechanisms."[2]

If the neocortex was to be exploited most efficiently as the new mammalian adaptation that stored, sorted, integrated, and recalled smell, sound, and vision maps in the dark, it could do so only when the brain was warm. So I presume that the neocortex would not have functioned properly when an animal slumped back from a warm endothermic state into the cooler, unregulated ectothermic state. Brain tissue uses more energy per unit of mass when at rest than does any other tissue. Gray matter is costly stuff. Large brains require more energy per unit of time, so more food per unit of time. The neocortex, for all its marvelous capacity, was an expensive innovation in the evolution of endothermy, yet, ultimately, it is the innovation that separated the mammals, and most certainly humans, from all other animals.

The third and last encephalization pulse occurred in true mammals. It involved a vastly improved sense of smell. Deep within the nasal cavity, after the inhaled air has been heated and humidified, it passes over the odorant receptor neurons in epithelial tissues attached to the cribriform plate. In true mammals the cribriform plate evolved from ossified turbinal bones. The plate separates the olfactory bulb of the brain from the nasal cavity. It is highly perforated with olfactory nerves that send their smell signals from the epithelium straight to the olfactory bulb. The evolution of the cribriform plate increased the surface area for olfactory receptors tenfold. This vastly improved the sense of smell and thus armed mammals with ways of finding food and reproducing in a dark world of aromas.

The increase in the size of the olfactory region and other parts of the brain doubled the encephalization quotient in true mammals. Today the echidnas have a quotient of about one, and the platypus about 0.75. Dogs and cats, for example, have quotients of more than one, whereas the pangolin, armadillo, and Virginia opossum have quotients less than that of *Hadrocodium*, that is, less than 0.50. These latter mammals, and many other small mammals with low quotients, have one thing in common: poor eyesight and/or defense mechanisms.

Whereas new innovations such as a warm neocortex and a good sense of smell would have conferred huge fitness benefits on mammaliaforms and true mammals, they were not necessarily "switched on" at all times. An important ongoing debate that I emphasize here concerns whether mammaliaforms had the newly evolved capacity to produce endogenous heat and maintain a continuous, constant body temperature. If a small mammaliaform could find a safe refuge in which to rest, safe from predators, why would it waste energy staying warm when there were no simultaneous fitness benefits to being warm? Remember, too, that being active increased the chances of being eaten by a predator. I emphasize this point because several theorists on the evolution of endothermy have emphatically denied that any departure from endothermy was the ancestral condition. To them, the evolution of endothermy was an all-or-nothing gradual shift toward a consistently elevated, regulated constant body temperature. Once an animal had attained an endothermic state, they claim, it stayed like that, throughout the twenty-four-hour cycle, throughout the year, and throughout the Jurassic and the Cretaceous, until the appearance of new high-latitude cold habitats in the Miocene, within the last 10 million years or so—nearly 200 million years of sustained constant body temperature. These theorists are arguing that the capacity of an endotherm to deviate from homeothermy and enter a state of torpor or hibernation is a highly *derived* adaptation that evolved in Late Cenozoic mammals only.

Tom Kemp from Oxford University, the author of the Correlated Progression Model of endothermy, is, in my interpretation, quite clear in his stance: "The various 'subendothermic' modern mammals can readily be shown by cladistic analysis to be secondarily specialized rather than relics of earlier evolutionary stages: neither daily torpor in bats, nor low metabolic rates of desert hedgehogs, nor the virtually ectothermic nature of the naked mole rat are reliable indicators of actual ancestral grades."[9]

In part I believe otherwise. I say "in part" because the naked mole rat (*Heterocephalus glaber*) did indeed undergo a metabolic reversal within Rodentia when the family it belongs to, the Bathyergidae, took to the underground realm in the Late Cenozoic. The Damara mole rat, the focus of my PhD research in the Kalahari Desert, did likewise. But "subendothermy"— which I presume to be synonymous with my definition of basoendothermy in bats and hedgehogs—is much more likely to be an ancient ancestral state. Gordon Grigg and his colleagues provided the most compelling early argument that torpor and hibernation were integral components of endothermy and represent the ancestral condition: "Noting that hibernation and torpor are almost certainly plesiomorphic (ancestral, primitive), and that heterothermy is very common among endotherms, we propose that homeothermic endothermy evolved via heterothermy, with the earliest protoendotherms being facultatively endothermic and retaining their ectothermic capacity for 'constitutional eurythermy.'"[8] A eurythermic animal is one that is capable of tolerating a wide range of body temperatures.

I agree. The source of the conceptual disparity between intermittent and sustained endothermy in early mammals has roots buried deep in a consequential Holarctic paradigm. The Holarctic is the biogeographical zone that combines the Palearctic and the Nearctic: North America and Europe.

Thomas Kuhn coined the term "paradigm" as universally recognized scientific achievements that, for a time, provide model problems and solutions for a *community* of practitioners. A Holarctic paradigm, then, solves model problems and solutions for Europeans and North Americans. Holarctic paradigms are *consequential*, I'd argue, when they are formulated from deep personal perceptions of living at high latitudes in highly seasonal environments—snowy, icy, sub-zero, gray environments in winter—life in the cold hard-wired into the childhood neocortex. I certainly do not mean to imply that they are malicious perceptions.

I have spent much of my career trying to convert Holarctic hibernation paradigmists to the idea that hibernation is a relict physiological condition from the Mesozoic and not a condition that evolved at high latitudes in the cold only. Allow me a slight digression here to tell you this story.

I generally do not attend as many scientific conferences as I should, because modern aviation to me is a vulgar experience. I restrict my attendance to meetings that focus on subjects that excite me to the point that the benefits of the meeting far outweigh the horrors of getting there and back. By far my favorite meetings are the international hibernation conferences held every four years. I have come to know well those who attend them regularly, and many are dear friends. They include Holarctic hibernation paradigmists, who have been responsible for having organized some of the meetings in the coldest places in the world, deliberately.

The first hibernation conference that I attended was held on the slopes of Cradle Mountain, Tasmania, in July 1996. Along with several other delegates, we were collected at the airport by what I was later informed was a "typical Tasmanian," driving a small bus. He was short and had an enormous paunch and a wild white beard. As we entered the national park near the top of Cradle Mountain, we spotted an increasing number of dead Tasmanian devils and other marsupials on the side of the road that wound its way through ancient King Billy pine trees (*Athrotaxis selaginoides*) to the lodge. One of the delegates went up to the driver and asked him how the animals had died. The driver sat upright, rigged up his public address system, and held an ancient studio-type microphone to his chest. He cleared his throat, making some disgusting sounds, and spat a gob out of the window.

"Ladies and gentlemen," said he, "them dead animals, for your information, is vermin—we aim for the bastards."

It dawned on me then that Australians who live in the cities are not quite the same as those who do not.

The conference was the tenth International Hibernation Symposium, and it was titled Adaptations to the Cold.

I was young, unknown, naive, and more than a bit opinionated. I presented a talk about my work with several colleagues titled "The Low Basal Metabolic Rates of Marsupials: The Influence of Torpor and Zoogeography."[13] I was trying to sell the Aussies an argument about their own dear marsupials. I tried to convince the symposium delegates that they should stop regarding torpor and hibernation as physiological responses to cold, highly seasonal environments only. I also tried to convince them that the majority of hibernators and daily heterotherms occurred outside the temperate zones, in the tropics and semi-tropics.

Stewart Nicol from the University of Tasmania gave a talk titled "Hibernation in the Echidna: Not an Adaptation to Cold?"[14] Delegates were uncomfortable that the two of us were rocking the Life-in-the-Cold boat, their beloved paradigm, which had survived forty-odd years to reach the tenth iteration of the symposium unscathed.

Another presentation at the conference added fuel to our skepticism of the Life-in-the-Cold paradigm. Jutta Schmid from the University of Tübingen and Sylvia Ortmann and her colleagues from Philipps University in Marburg, Germany, showed that the smallest primate on earth, the mouse lemur (*Microcebus*) from Madagascar, used daily torpor. The significance of these data seemed merely to bounce off most Life-in-the-Cold paradigmists. They failed to recognize that a small primate in the semi-tropics that used torpor on a daily basis posed a threat to their long-cherished ideas.

The next conference was held in a place that in winter is extremely cold: Jungholz, Austria, a ski resort. Mercifully, it was held in midsummer, and the climate was delightful. In Jungholz I presented a paper titled "Daily Heterothermy in Mammals: Coping with Unpredictable Environments." Gordon Grigg presented a paper titled "Hibernation by Echidnas in Mild Climates: Hints About the Evolution of Endothermy?" Kathrin Dausmann from the University of Hamburg provided new and even more stunning data from Madagascar: she reported true multiday hibernation in dwarf lemurs (*Cheirogaleus*). Dausmann showed that some primates hibernate in the semi-tropics. Her paper was titled "Body Temperature and Metabolic Rate of a Hibernating Primate in Madagascar: Preliminary Results from a Field Study." The case against the Life-in-the-Cold paradigm was growing.

Four years after Jungholz, the twelfth conference was held over seven days on board the Holland America cruise ship MS *Veendam*, which sailed from Vancouver, Canada, to Seward, Alaska. It was very cold, but the glaciers we slid past were beautiful, as were the sea otters and puffins. The conference was called "Life in the Cold: Evolution, Mechanisms, Adaptations, and Application." I must admit that the parties on board every night were highly memorable, but the conference itself was not. I do remember making a pitch, together with my former PhD student Andrew McKechnie, for the next conference, and winning, against a pitch for the next meeting to be held Stockholm.

Andrew and I decided to organize the next conference as far away from the cold, high latitudes as we could. We chose a very non-Life-in-the-Cold venue: Swakopmund, Namibia, in the Namib Desert. It was time, we reckoned, for the Life-in-the-Cold paradigm, at least as the conference series titles were concerned, to be put to bed. Our conference was called "Hypo-metabolism in Animals: Hibernation, Torpor, and Cryobiology."[15] Hypo-metabolism simply means a lower-than-normal metabolic rate. Andrew set a precedent onboard by giving the first talk on torpor in birds; mammal research had dominated precious conferences.

It is a central thesis of this book that endothermy evolved in the tropics. The thermoregulatory phase—that is, the stage of endothermy in which a high per-gram metabolic rate evolved to maintain a constant elevated body temperature in small nocturnal, dogtooth-like proto-mammals—also oc-curred in the tropics. So the ultimate jump from ectothermy to endothermy in mammals took place in the tropics, at night. Let's get back now to what really happened in the dark Jurassic forests.

About 165 million years ago, the otter-like mammaliaform *Castorocauda lutrasimilis*, which weighed between 500 and 800 g (17.6–28.2 oz), took to hunting and feeding on fish. We have deduced this information from a superb Chinese fossil (shown above in Figure 7.1), one of the most impor-tant to have ever been discovered, because it shows the first unquestionable imprints of fur in a mammal. *Castorocauda* had webbed feet and lived in burrows on steam banks. It also had a broad and flattened beaver-like tail and was probably very agile in the water, as is a modern otter. Its teeth were much like those seen today in mammals that eat fish. But *Castorocauda* did not persist into the Cretaceous. Like so many other early mammaliaform experiments, it perished in the Late Jurassic, along with *Morganucodon*.

Castorocauda was a dicodont (Dicodontidae) that evolved from the mor-ganucodonts. The dicodonts, in turn, gave rise to the ancestor of the first real mammal, which left two descendants, one that gave rise to the egg-laying mammals, the monotremes, and the other that gave rise to all other mammals, among them the living placental and marsupial mammals. In between the monotremes and the other crown mammals, however, the Ju-rassic diversification continued unabated in the evolution of the eutricon-odonts, multituberculates, spalacotheriids, cladotherians, and boroespeni-dans. With the exception of the multituberculates, these mammal families went extinct at or near the Cretaceous-Cenozoic boundary. Don't worry about these names now; I'll introduce them again shortly.

At the onset of the Jurassic the stage was set for an explosion of diversity fueled by the newly acquired capacity for keeping a small body warm at night in many different environments, especially during the breeding season. Despite the supposed persistent pressure placed on the mammals by dinosaurs during the dark nights of the Jurassic and Cretaceous, they did not skulk away for 150 million years and remain tiny insect eaters that lived on the forest floor and in the trees. Advanced endothermy equipped them with the capacity to diversify and occupy niches that larger dinosaurs could not exploit. The mammals underwent a spectacular explosion of new forms in the dark, forms that gave rise to the lineages of the modern marsupials and placental mammals. They exploited new terrestrial, arboreal (living in trees), aquatic, and underground environments,[16] and they even took to the air.[17] In a dinosaur-dominated world, they owed their persistence heavily to endothermy, their small size, and their fingers and toes.

The earliest non-monotreme mammal was *Fruitafossor*, discovered not in China but in Fruita, Colorado—hence its name. The "fossor" part alludes to its burrowing or fossorial lifestyle. *Fruitafossor* was a termite specialist, much like today's anteaters, armadillos, aardvarks, pangolins, and numbats. It was the size of a large mouse (49 g) (1.7 oz), had small eyes, and had enormously powerful front limbs that it used for burrowing into termite mounds. It persisted for most of the Jurassic, but, along with *Castorocauda*, went extinct during the Late Jurassic.

Fruitafossor was the first true non-monotreme mammal. It was the ancestor of the eutriconodonts, a group of mammals that diversified from the Late Jurassic and throughout the Cretaceous. The group includes a creature that was perhaps the biggest success story of the Mesozoic—that is, if you enjoy revenge stories. *Repenomamus robustus* was the biggest and meanest of all extinct Cretaceous mammals. It was a 12-kg (26-lb) carnivore with large incisors and very powerful jaw muscles, very different from the other small, mostly insectivorous mammals of the time. *Repenomamus* was the only mammalian dinosaur basher that ever evolved (Figure 9.3). It specialized in hunting and feasting on baby dinosaurs. Yaoming Hu and his colleagues from the Chinese Academy of Sciences and the American Museum of Natural History described a fossil of *Repenomamus* with a baby *Psittacosaurus* dinosaur in its stomach: "The juvenile skeleton of *Psittacosaurus*

Figure 9.3. Reconstruction of the dinosaur-hunting *Repenomamus robustus* with a hatchling of the dinosaur *Psittacosaurus* in its mouth. (Image by and used with the kind permission of Nobu Tamura, CC BY 3.0, http://ntamura.deviantart.com)

is the remaining stomach content of the mammal. The head-body length of the juvenile *Psittacosaurus* is estimated to be 140 mm (5.5 in), about one third of the head-body length of the *R. robustus*. . . . There are at least seven teeth on each jaw quadrant of the juvenile *Psittacosaurus*, most of which are worn. This demonstrates that the *Psittacosaurus* skeleton is not from an embryo. A few long bones are preserved in articulation . . . suggesting that the juvenile *Psittacosaurus* was dismembered and swallowed as chunks."[18]

It is gratifying to know that some mammals did manage to tyrannize dinosaurs. *Psittacosaurus* was a bipedal ornithischian herbivorous dinosaur with a large, hooked, parrot-like beak; hence its name. It was about the size of a Labrador dog and weighed more than 20 kg (44 lb). It sported tubular bristles on the upper surface of its tail that probably served a communication function. *Repenomamus* was not exactly a highly mobile carnivore, but it was endothermic, so it could hunt at night. It probably hunted by snatching babies from the periphery of their parents' protection and waddling off into the dark to rip the body into swallowable bits.

Some early mammals closely related to the eutriconodonts were the first mammals to take to the air, gliding over large distances cheaply and evading predators. *Volaticotherium antiquus* is known from a Chinese fossil of a glider that weighed about 70 g (2.5 oz) and had a patagium—a flying membrane of fur-covered skin stretched between its arms and its legs. It lived at the same time as *Castorocauda*, about 164 million years ago. Although it glided very much like a modern flying squirrel, unlike the squirrel, it ate insects. *Volaticotherium* also did not survive the end of the Jurassic.

The abundance of good-quality Jurassic and Cretaceous fossils from China sometimes creates a lopsided impression about the distribution of extinct mammaliaforms and true mammals on earth. They were most certainly not restricted to China. Fossils of *Argentoconodon*, for example, show that there were gliding mammals in South America as well.[19]

The Multituberculata were a highly successful group of now extinct rodent-like mammals. They were the only mammals to survive the meteorite impact at the end of the Cretaceous along with the living mammals: the monotremes, marsupials, and placental mammals. They went through a massive diversification starting 20 million years before the extinction event,[20] a time when the dinosaurs were still kings of the castle. Once the nonavian dinosaurs went extinct, they persisted well into the Cenozoic, until the Late Eocene, 35 million years ago, when the forests gave way to open landscapes. At this time the modern rodents were emerging and undoubtedly provided stiff competition for food. The multituberculates persisted for 120 million years, longer than any other mammal group. They managed to compete very successfully with the dinosaurs, but could not cope with a mammal, the true rodent, which had essentially evolved independently to have some of the best characteristics of the multituberculates, with refinements.

The multituberculates had highly derived teeth and had a diverse diet. There were herbivores, frugivores, granivores, root and bark eaters, egg eaters, insectivores, carnivores, and omnivores. The complexity of their teeth was produced by an increase in the numbers of cusps on the molars— hence the name Multituberculata—and the relative size of the molars. The increase in dental complexity was also matched by an increase in the average body size during the Late Cretaceous and Early Paleogene. The biggest was *Bubodens magnus*, which reached an estimated 5.25 kg (12 lb). This animal was also herbivorous, which is not surprising, because one of the themes that I emphasize throughout this book is that successful herbivory in endotherms demands an increase in body size to maximize fermentation efficiency.

Gregory Wilson from the University of Washington and his colleagues believe that the unexpected explosion in ecological diversity and body size of the multituberculates at a time when other mammals were still under severe pressure from the dinosaurs can be attributed to the evolution of the flowering, seed-producing plants, the angiosperms.[20] They proposed that many of the earliest angiosperms were herbaceous, nonwoody plants in which the nasty compounds that deter herbivores today, secondary compounds, had yet to evolve. Multituberculates, it seems, were able to pounce on the angiosperm food source before any other animal of mammalian, reptilian, or dinosaur lineage.

The ancestors of the modern mammals managed to survive the holocaust at the end of the Cretaceous. The Multituberculata managed to survive as well, undoubtedly for the same reasons: they could resort to their ancestral ectothermic physiological state and enter a state of hibernation for long periods of time. The mammal *Bauplan* had finally come together. A small body size, a nocturnal active phase, thermoregulation, torpor, hibernation, diet diversification, fingers and toes, new forms of locomotion, and the occupation of thermally comfortable hideaways during the daytime—these were crucial characteristics for survival in Late Mesozoic and Early Cenozoic mammals.

PART TWO

The Story in the Bloods

TEN

The Day of Reckoning

In 1980, the same year that I was booted out of the University of Cape Town (UCT) for being unable to pay my fees as an undergraduate, a momentous paper was published in *Science* by a team of scientists led by a father and his son (Figure 10.1). Luis Alvarez, a nuclear physicist, and his son Walter, a geologist, provided the best explanation yet for the extinction of the dinosaurs.[1] They argued that the dinosaurs were killed by a meteorite strike, and nearly four decades later, their claim is still gaining ground and scientific support. I do not know of any scientist who has managed to pick a hole in this decades-old paradigm.

Luis was adventurous. He flew in the B-29 Superfortress bomber that accompanied the *Enola Gay* in the bombing of Hiroshima on August 6, 1945. He was also smart; he won the Nobel Prize in Physics in 1968 for his contributions to particle physics.

Ever since the 1980 publication, biologists have had to think very differently about life on earth. I returned to UCT in 1981 and, being "all ears" at the time, was beguiled by the simplicity of the Alvarez team's argument. At the limestone boundary between the Cretaceous and the Cenozoic, Walter collected samples and took them to his father's laboratory at the University of California, Berkeley. After analysis by two colleagues, the team found high concentrations of an element, iridium, that occurs naturally on the earth's surface in very low concentrations. Iridium is such a dense material that most of what did occur on the earth's surface at one time sank to the

Figure 10.1. Luis Alvarez (left) and his son Walter. Walter's right hand is resting on the top of the Cretaceous, whereas Luis has his hand on the Early Cenozoic. The photo was taken at a limestone outcrop in Gubbio, Italy, from which Walter collected the original samples that showed high concentrations of iridium at the Cretaceous-Cenozoic boundary. (Image in the public domain, courtesy of the U.S. Government, https://en.wikipedia.org/wiki/File:LWA_with _Walt.JPG#/media/File:LWA_with_Walt .JPG)

core. In extraterrestrial bodies, however, such as meteorites, iridium occurs in high concentrations.

The Alvarez team linked the extraterrestrial iridium at the Cretaceous-Cenozoic boundary to the extinction of the dinosaurs. Their story solved one of the biggest puzzles in paleontology: the sudden disappearance of the dinosaurs at the boundary. Dinosaur fossils are abundant below the boundary, but above it, they vanish.

At the end of the Cretaceous period, 66 million years ago, the world was a warm, moist, steamy place. The great Gondwana continent had yet to finish splitting up into the pieces that we know today as South America, Africa, Madagascar, India, Australia, and Antarctica. But then a meteorite the size of Manhattan, 10 km (6.2 mi) wide, slammed into the planet at Chicxulub, on the Yucatán Peninsula of Mexico. It was traveling at 30,000 km/h (18,641 mi/h). The impact left a crater 200 km (124 mi) wide that can still be seen from space today. Luis Alvarez would not have survived the observation of *this* explosion—it released more than a billion times more energy than the bomb that was dropped on Hiroshima. It would have incinerated his B-29 Superfortress, *The Great Artiste*.

The immediate effects of the impact were dramatic. All land surrounding the Gulf of Mexico was flooded by a massive tsunami. Although the meteorite hurtled into fairly deep water, material from the seabed was blown out by the impact with such force that it exited the earth's atmosphere—it was blown into orbit. The seabed was composed of gypsum—that is, calcium sulfate—so great clouds of sulfur dioxide were generated. When the material returned through the earth's dense atmosphere it was superheated into quartz. These quartz particles and iridium can be found everywhere on earth at the Cretaceous-Cenozoic boundary layer. The sulfur dioxide descended as acid rain, which reduced the amount of sunlight penetrating through the atmosphere to the earth's surface; the acid rain and lack of sunlight together killed all plants. The whole planet was affected by the meteorite's impact.

The radiation from the ballistic reentry of material blasted out by the meteorite's impact heated the atmosphere severely for about three hours following the impact. Areas within 1,000 km (620 mi) of the impact were burned to a cinder. Other parts of the globe suffered a blast of infrared radiation sufficiently intense to ignite wildfires where there was fuel—in grasslands, woodlands, and forests. Only heavily cloud-covered regions of the earth would have been spared the firestorms. These were the initial short-term effects of the impact.

The long-term effects were far more devastating to life on earth. The biggest long-term problem was high-altitude dust that blocked out sunlight. The world became a dark, cool place, on average about 10°C (18°F) cooler than before the meteorite's impact. The darkness lasted for at least a year, some argue, perhaps even for as long as ten years. Any organism that relied directly or indirectly on sunlight to sustain its life was in serious trouble. Light is essential for photosynthesis by plants on land and by phytoplankton and other organisms in the sea. Sunlight is also essential for fixing carbon dioxide into the sugars from which food for all animals is built. The carbon dioxide–fixing organisms died, and along with them the herbivores: dinosaurs, reptiles, fish, and insects. Animals that fed on herbivores died, too: carnivorous dinosaurs, predatory fish, and insects. The Cretaceous-Cenozoic event was the second biggest extinction event ever to have occurred on earth, second only to the Permo-Triassic Mass Extinction. And, yet again, as in the case of the Permo-Triassic event, the ancestors of the mammals survived, and so did the ancestral marsupials, egg-laying monotremes, the Multituberculata, and the neornithine birds. On the ectothermic side, many fishes, amphibians, and reptiles (snakes, lizards, turtles, and aquatic crocodiles) also survived. How did they manage to survive the fire and brimstone when so many other vertebrate groups did not?

Douglas Robertson from the University of Colorado and his colleagues have undertaken research on the conditions on earth during the first hours after the meteorite's impact.[2] Any animal that was deeper than 10 cm (3.9 in) underground would have been completely unaffected by the initial infrared blast from the impact. In the seas, lakes, and streams, a few mm (0.08 in) of water would have protected aquatic animals, because most of the infrared heat from the blast would have been dissipated at the surface as the heat of vaporization released during the evaporation of water.

The three mammal lineages that survived the Cretaceous-Cenozoic extinction had characteristics in common. They were small, less than 1 kg (2.2 lb). They could therefore easily enter burrows underground to escape the scorching first hours of the impact. Large dinosaurs could not do this. Any animal that could not hide underground or periodically duck underwater would have died by roasting. The mammal ancestors were insectivorous. They ate things like earthworms, pillbugs, and cockroaches, creatures that can survive for long periods of time in the dark without reliance upon sunlight-dependent foods. The food supply of the ancestral mammals would therefore not have been severely affected by the long-term effects of the impact. It is a great irony that the very features that persisted in the mammal lineage because of the spectacular rise to dominance of the dinosaurs in the Triassic — that is, small body size and having fingers and toes — were perhaps the important features that spared them from extinction at the Cretaceous-Cenozoic boundary.

Tai Kubo from the University of Tokyo has argued that the loss of the flatfoot in dinosaurs prevented them from evolving to body sizes smaller than 1 kg (2.2 lb).[3] Once the digitigrade foot evolved in the bird lineage, it allowed the dinosaurs to become very large, to hog the large-body-size niches of the Mesozoic, and to dominate the smaller flatfoot mammals. But this dominance spelled their downfall at the Cretaceous-Cenozoic boundary because, except for the small neornithine birds, they were too big to hide from the inferno of the infrared blast following the meteorite's impact.

The name of your and my placental ancestor is Schrëwdinger. It was small, about the size of a rat (Figure 10.2). It looked like a possum but was not a possum, because possums are marsupials. Schrëwdinger was a placental mammal—the first placental mammal. It had a long, furry tail, and it ate cockroaches, worms, and all things buglike. Schrëwdinger cannot have a scientific genus and species name because there is no fossil of Schrëwdinger, no "type specimen."

Schrëwdinger is the product of the largest reconstructive study of an ancestral creature ever undertaken.[4] It is a virtual mammal, a virtual ancestor. It was bestowed with its name in an Internet naming competition. The name Schrëwdinger finally won out over the leading contender for a while, "Ralph." It was named after Erwin Schrödinger, the Austrian physicist made famous by his thought experiment called Schrödinger's cat, in which a cat is simultaneously both dead and alive, in a state known in quantum physics as quantum superposition. The different spelling of the animal's name comes from its shrew-like appearance.

Schrëwdinger's reconstruction was a massive team effort by paleontologists and molecular geneticists from universities and museums in New York, Tennessee, Pomona, Pittsburgh, and Louisville and in Stony Brook, Florida, and New Haven, Connecticut, in the United States and in Toronto in Canada. The senior author of the 2013 publication was Maureen O'Leary from the School of Medicine at Stony Brook. The findings of the study stunned many biologists, including me, mostly because the obvious seemed to have been confirmed, finally.[4]

The O'Leary study used 4,541 different measurements of the dimensions of the bones of both extinct and living mammals and pooled these data with the molecular sequences of twenty-seven nuclear genes of living mammals. These data were then entered into a computer program written specifically for the project, which reconstructed the family tree of the placental

Figure 10.2. A reconstruction by Carl Buell of the ancestral placental mammal that survived the meteorite impact at Chicxulub, Mexico, 66 million years ago. Known informally as Schrëwdinger, it is named after Schrödinger's cat trilogy, in which a cat is simultaneously dead and alive in a state known as quantum superposition. Schrëwdinger survived the catastrophe of the meteorite strike at the end of the Cretaceous, probably because it could hibernate for long periods of time. (From O'Leary, M. A., et al., "The Placental Mammal Ancestor and the Post-K-Pg Radiation of Placentals," *Science* 339 [2013]: 662–667; printed with the kind permission of the American Association for the Advancement of Science [License no. 3958811323514])

mammals. At the base of the tree was a single ancestor, for which there is no fossil. The combined data were used to reconstruct the ancestor with some reliability. That ancestor is Schrëwdinger.

Several conclusions were reached by this study that will have paleontologists and molecular geneticists scratching their heads and arguing for decades. Indeed, the arguments have already started, but I am not going to venture there. For the purposes of my story, I am sticking to Schrëwdinger and the date when it is thought to have first appeared. The main claim of the O'Leary study is that the modern placental lineages evolved *after* the dinosaurs were toasted at the Cretaceous-Cenozoic boundary. This is called the "short-fuse model." This claim is not new. For a long time it has been argued that the dinosaurs suppressed the evolution of the placental mammals during the Jurassic and the Cretaceous, and that it was only after the dinosaurs went extinct that the mammals could lay claim to the planet. It is a nice, neat, simple idea, a perfect induction.

Molecular geneticists dislike the O'Leary claim because their conjectures about the timing of the origin of the placental mammals have always placed it much earlier than what the fossils say. Their times of origin are always well within the Cretaceous, much older than 66 million years. This is called the "long-fuse model." Molecular biologists use molecular clocks to date when splits occurred in family trees. After measuring the similarity of gene sequences in living mammals, they make assumptions about how rapidly the mutations responsible for sequence differences occurred. They can then group species together based upon sequence similarity and can also estimate when clades split using their estimates of the rate at which mutations occur. They calibrate their molecular clocks against fossils of reliably known ages.

There is an important evolutionary ramification that emerges from the argument about whether placental mammals' evolution followed the short- or the long-fuse model. The O'Leary study claimed that all of the major placental lineages—those of the rabbits, rodents, primates, tree shrews, flying lemurs, bats, insectivores, carnivores, and ungulates—evolved within several hundred thousand years of the Cretaceous-Cenozoic boundary. These data suggest a remarkably fast increase in the rate of evolution, a Paleocene explosion of new mammalian forms that made the Jurassic radiations look trifling. The onset of the Cenozoic was pulsed with evolutionary innovations, some of which have lasted until now, but many of which lasted only until the warm phase of the Cenozoic was over, let's say up to around 45 million years ago. Molecular geneticists don't like sudden pulses—it makes their molecular clocks look wonky and unreliable; good clocks should not speed up and slow down.

Before we discuss how Schrëwdinger managed to survive the Cretaceous-Cenozoic extinction event, let's see how the birds managed it as well.

Birds evolved from dinosaurs, yet all dinosaurs went extinct and the neornithine birds did not. However, an even larger number of ancestral birds, the Enantiornithes, did not survive the impact.

Douglas Robertson and his colleagues argued that aquatic birds would have survived the meteorite impact not because they could dive but because they nested in burrows or tree holes and rocky cavities. A reconstruction of the ancestral neornithine bird, the avian Schrëwdinger counterpart, has yet to be undertaken, so we do not know what it looked like. But we do know that the ancestors of several bird lineages survived, just as in the case of the mammals, in which the ancestors of the monotremes, marsupials, and placental mammals persisted. However, we do not know how big any of these ancestral birds might have been or whether they burrowed or were aquatic or terrestrial.

Julia Clarke from North Carolina State University and her colleagues have described a Late Cretaceous fossil bird, *Vegavis iaai*, from Vega Island, Western Antarctica, which, they argue, is most closely related to the Anseriformes, that is, to the ducks (shown in Figure A.5).[5] If this is correct, it is one of several modern bird lineages that must have lived side by side with the dinosaurs shortly before the meteorite impact.

The biggest genomic study on birds to date, published in *Science* and authored by dozens of scientists from all over the world, has cast more light on what happened at the Cretaceous-Cenozoic boundary.[6] Senior author Erich Jarvis from Duke University and his colleagues show that it is fairly certain that the ancestral ratites (Palaeognathae) and tinamous (Tinamidae) made it across the boundary. These two families are estimated to have diverged 84 million years ago, that is, 19 million years before the meteorite impact. The chickens (Galliformes) and the ducks (Anseriformes) diverged very soon after the boundary, so their ancestor made it as well. That makes three definite survivors so far. But perhaps the most intriguing question is whether a single ancestor of all other birds—that is, the ancestor of Neoaves—survived as well or whether the Neoaves ancestor had already diverged into the two massive clades within Neoaves, the Columbea and the Passerea—identified in the Jarvis study—before the meteorite impact.

The ostriches and other ratites branched off first within the Palaeognathae, the oldest of the bird groups that survived the Cretaceous-Cenozoic extinction. The oldest ratite fossil is *Remiornis heberti*, which roamed Europe during the Late Paleocene, about 57 million years ago. It was not a small bird, and neither was another flightless bird of the time, the huge, herbivorous *Gastornis*, which had a colossal beak and was closely related to the ducks. So without Early Paleocene fossils from the nine-million-year gap between the Cretaceous-Cenozoic boundary and the *Remiornis* fossil, we do not know how big the ancestral ratite was that survived the Cretaceous-Cenozoic boundary. But I think we can make some educated guesses.

The tinamous from South America are the sister clade to the ratites. The living species can fly; they are ground-dwelling birds about the size of a partridge. If we reasonably assume that flight evolved only in the birds, the capacity for flight in tinamous has important evolutionary implications because it strongly suggests that the ancestral ratites could fly. Indeed, the current opinion is that the ancestors of the flightless ratites, the ostriches, rheas, kiwis, and emus, got to their respective southern landmasses by flying and then subsequently became flightless. If this was the case, we might expect the ancestral flying ratite to have been much smaller than its modern descendants. We know that the bird miniaturization process occurred prior to the extinction event and that, on average, birds that used muscle-powered flapping flight initially did not weigh more than a kilogram. Small ancestral birds could fly, but big ones could not. Moreover, small birds were much more likely to have survived the Cretaceous–Cenozoic boundary by sheltering, for example, in rocky crevices, than were large, exposed, flightless birds the size of ostriches.

I have done some yet unpublished body-size reconstruction exercises with my students using maximum-likelihood approaches to estimate the body size of the ancestral neornithine bird. Using various models of evolution we obtained estimates that it weighed less than 200 g (7 oz). Our estimate suggests that the ancestral bird was smaller than *Archaeopteryx*; it was about the size of a pigeon.

The swamp-dwelling screamers, which are a sort of blend between ducks and partridges, along with the ducks, are sister clades that are closely related to the chicken lineage. So it would seem that the ancestor of these three lineages, that is, the ancestral galloanseriform, was a flying, ground-dwelling, water bird. It was probably a shorebird of sorts. Modern screamers inhabit marshy areas and have partially webbed feet. Moreover, their chicks are aquatic.

Ducks are ducks. They were, and still are, aquatic. So by Douglas Robertson's argument, it would be fairly easy to explain how the ducks and the screamers survived the immediate infrared blast from the meteorite impact, but we'd need to invoke the same explanation that we used for the ratites and the tinamous to explain how the chickens made it across the Cretaceous-Cenozoic boundary without becoming the Kentucky variety.

What was the ancestor of the Neoaves like? The closest relatives today, according to the Jarvis study, are the grebes, flamingos, doves, sandgrouse, and mesites within Columbea and a massive radiation of thirty-three other bird families within Passerea. The explosive radiation, both within Columbea and Passerea, mirrors exactly what happened in the mammals following the Cretaceous-Cenozoic extinction event. Again, within a staggeringly short period of evolutionary time, a multitude of new bird groups evolved, and by 50 million years ago, all crown orders within Neoaves had made their appearance. Unlike the O'Leary study on mammals, though, the Jarvis study on birds did not pool genomic data with morphometric data—that is, data on bone measurements—to try to reconstruct the ancestral bird. To me, it would be particularly intriguing to know what this bird was like because it could be considered the avian Schrëwdinger counterpart. It would be the avian counterpart because it was the ancestor of the biggest family tree of birds, just as Schrëwdinger was the ancestor of the placentals, which have the biggest family tree in the mammal phylogeny.

It is a compelling question whether or not the ancestors of Neoaves, the ducks and the ratites, might have relied on hibernation to survive the long-term effects of the meteorite impact. Today ratites, tinamous, screamers, and ducks do not hibernate.

As I was contemplating this issue, Richard Prum from Yale University and his colleagues published a bird phylogeny that they claim is an improvement of the Jarvis tree.[7] They, too, believe that the Palaeognathae and the Galloanserae evolved before the Cretaceous-Cenozoic boundary and that their ancestors managed to survive the bolide impact, but what made their phylogeny exciting to me is that the group of birds that branched off first in their Neoaves were the Caprimulgiformes, the poorwills or night-jars, swifts, and hummingbirds within a clade called the Strisores. As I will discuss later, poorwills are the only modern birds in which hibernation is known, although many other birds, especially hummingbirds, use daily torpor on a highly regular basis.[8] Thus an argument could be developed that the genes for long-term hibernation were present in the ancestral Neoaves but were lost in all lineages more derived than the Caprimulgiformes.

To pursue this question further we need to look for the first evidence of endothermy, and deviations from endothermy, in the archaic birds—that is, in the Late Cretaceous birds, those that lived before the Chicxulub impact.

The bird lineage consisted of two major groups by the Late Cretaceous, the enantiornithines and the Ornithurae (shown in Figure A.5). All members of the diverse Late Cretaceous enantiornithines went extinct at the Cretaceous-Cenozoic boundary, but some members of the ornithine birds got through the disaster and went on to diverge into modern birds. But why did not a single species of enantiornithine survive when at least four ornithine lineages did?

Anyusuya Chinsamy from the University of Cape Town, with several collaborators, published a study that examined the bone histology of an enantiornithine bird and one of the oldest ornithine birds, *Patagopteryx deferrariisi*. They wrote: "The implication of true zonal bone in Maastrichtian

[Late Cretaceous] birds is that even half way through their evolutionary history, birds still lacked a typical avian physiology. If the LAGs [lines of arrested growth] in these birds mark true zonal bone and are a usual and natural phenomenon in these Late Cretaceous non-ornithurine birds, then this would imply that birds evolved endothermy and high sustained metabolic rates along their own evolutionary lines at a significantly later time in their history than has previously been proposed."[9]

What Chinsamy showed was that the enantiornithine bird had many lines of arrested growth in the bone. It periodically stopped growing, probably because of poor seasonal resource availability. The presence of lines of arrested growth means not that it hibernated but merely that it stopped growing at certain times of the year. But *Patagopteryx deferrariisi* had no lines of arrested growth at all, which indicates a stricter control of body temperature and metabolism. *Patagopteryx* was fully endothermic, so modern birds were definitely endothermic before the dinosaurs went extinct.

Patagopteryx deferrariisi had become secondarily flightless and was about the size of a chicken. So it was a relatively large bird compared with a poorwill or a hummingbird, or an enantiornithine bird. Both birds and mammals (except bears) lost the capacity for hibernation and torpor as they grew bigger, because they no longer needed it.[10] So *Patagopteryx* is not useful for inferring anything about ancient hibernation in Cretaceous birds.

No hibernation biologist would ever buy the argument that hibernation evolved independently in poorwills only and never existed in the archaic birds. If poorwills can hibernate, hibernation must have existed somewhere in its ancestry. The complexity of biochemical networks required for hibernation is simply too great for it to have suddenly appeared in one clade of birds only. The most likely explanation is that some ancestral birds were hibernators and that hibernation is an ancient characteristic of the avian lineage that was probably employed quite extensively during the Cretaceous. If we accept the preceding argument, we can propose that our small-sized ancestral ratite, tinamou, screamer, and duck, may have had the capacity for some or other long-term reduction in metabolic rate and body temperature. This capacity would have profoundly increased their chances of surviving the long-term ecological horrors of the Chicxulub impact.

Perhaps one explanation of why few modern small birds do not hibernate is that their ancestors became larger during the Paleocene and may have lost the capacity as they got bigger, as occurred in mammals.[11] For example, the next group of birds that evolved after the Caprimulgiformes, the Columbaves, includes not only the doves, but also the largest and heaviest modern flying bird, the Kori bustard *Ardeotis kori*, the males of which can attain 20 kg (44 lb). No endotherm today, mammal or bird, with a mass greater than that of a big male Kori bustard can fly. Birds seem to have gotten bigger and remained so throughout the Paleocene. Only in the Eocene did small body sizes reappear, for example, in the passerine birds. This hypothesis can be tested easily using modern methods of character state reconstruction.

Whereas the ornithine ancestors were primarily shorebirds, the enantiornithine birds, which were by far the dominant and most diverse birds of the Mesozoic, were all terrestrial forest dwellers. It is perhaps this difference that accounted for which clade went extinct and which did not. The forests, and hence the enantiornithine birds, were probably more adversely affected by both the short- and the long-term effects of the meteorite impact than were the other birds.

Soon after World War II, it was reported that poorwills, *Phalaenoptilus nuttallii*, were found hibernating in "crypts," shallow indentations on the surface of large granite boulders in the Chuckwalla Mountains in the Colorado Desert in California.[12] Poorwills are strictly nocturnal birds that feed on insects on the wing. In the late 1990s, Mark Brigham, University of Regina, Canada, sent his PhD student Christopher Woods to the Sonoran Desert in southern Arizona to measure hibernation in poorwills in the field. It took several years to achieve, but Woods succeeded in obtaining skin temperature data from poorwills that he shaded from direct sunlight at their roosts to avoid passive heating during hibernation. Captured birds were fitted with harness-mounted temperature-sensitive radio transmitters of the type whose signal frequency changed with the bird's skin temperature. Brigham sent me the student's PhD thesis, and I digitized the data from one of the figures showing a six-week trace of skin temperatures (Figure 10.3). The data show

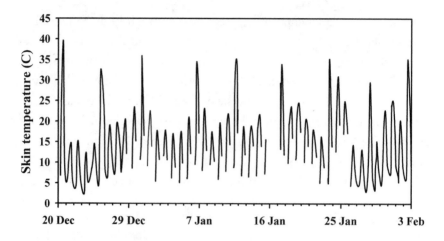

Figure 10.3. The skin temperature of a single common poorwill, *Phalaenoptilus nuttallii*, in southern Arizona that was continuously shaded experimentally for forty-five days to prevent passive solar warming at its roost. The bird remained inactive throughout that period. The temperature spikes in the graph show arousals from hibernation. Note that the skin temperatures are typically lower than internal body temperatures when the birds are at rest. (Data digitized and replotted from Woods, C. P., *Ecological Aspects of Torpor Use and Inactivity During Winter by Common Poorwills*, PhD thesis, University of Regina, Saskatchewan, 191 pages, 2002)

hibernation bouts lasting for five to seven days. These are the only hibernation data for birds that exist.

Since then hibernation has been confirmed in numerous North American caprimulgids. The Caprimulgidae are found on all continents except Antarctica, so it was interesting to explore the question about whether hibernation is used by poorwills that live outside North America as well.

Andrew McKechnie, my former PhD student, started to pursue this question when he found evidence of the use of daily torpor in the freckled nightjar, *Caprimulgus tristigma*, in the savannah habitat of northern South Africa.[13] These birds entered torpor before midnight and used the heat of the air the following morning to reheat. Andrew and his collaborators argued that what they had measured did not appear to be true hibernation like that which occurs in poorwills, but they suggested that it might be fruitful to investigate desert-dwelling nightjars in Africa. And herein lies a short story, about Ben Smit and Namaqualand.

Ben can manipulate nature. He can catch things that other humans cannot, especially birds, with his bare hands. He sings to the birds. He calls them, and they come to him, and then he catches them. His imitations of bird calls are stunningly real. I cannot tell the difference between his calls and the real thing, and he has fooled me on several occasions during our times together, sadly all too rarely in the field.

Ben set out to study hibernation in nightjars in a mountainous region of Namaqualand. He caught the birds, at night, by listening to their calls, creeping up on them, blinding them with a flashlight, and then pouncing on them. He used the same technique that Christopher Woods used on poorwills, fitting the nightjars with temperature-sensitive radio transmitters (Figure 10.4). That was the easy part of Ben's project. He then had to locate the birds in the mountains every night by tracking the transmitter radio signals from each bird. Once he got a signal he would record the frequency of the pulses to calculate the bird's skin temperature. The task is easy when

Figure 10.4. A freckled nightjar, *Caprimulgus tristigma*, in torpor in Namaqualand, South Africa. At the time that the photograph was taken, the bird's skin temperature was 10°C (50°F) measured with a collar-mounted, temperature-sensitive radio transmitter telemeter; the aerial of the transmitter is pointing backward over the bird's back. (Photo by and used courtesy of Ben Smit, Rhodes University, Grahamstown, South Africa)

you have only one bird collared. It gets exhausting when you have six, and only a precocious birdwatcher like Ben could have managed to find the birds every night and day and successfully obtained the data.

Ben's data were stunning.[14] When there is no lunar illumination and hence it is too dark, nightjars do not fly to catch insects on the wing (Figure 10.5). Instead they enter deep torpor on a daily basis. Their patterns

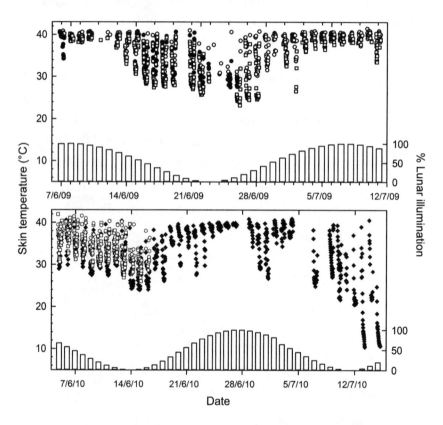

Figure 10.5. Data for the freckled nightjar, *Caprimulgus tristigma*, from Namaqualand, South Africa. The different symbols represent individual birds. The vertical bars at the base of each graph show the percentage of lunar illumination. When the illumination is low, the nightjars use torpor and allow their skin temperatures to decrease as low as 10°C in some birds. Skin temperatures are generally about 2°C lower than the core body temperatures. These data illustrate beautifully the flexible nature of endothermy relative to foraging opportunity. (Figure redrawn from Smit, B., Boyles, J. G., Brigham, R. M., and McKechnie, A. E., "Torpor in Dark Times: Patterns of Heterothermy Are Associated with the Lunar Cycle in a Nocturnal Bird," *Journal of Biological Rhythms* 26 [2011]: 241–248)

of torpor match the lunar cycles beautifully. The birds show the least use of torpor during the full-moon phase and the most during the new-moon or waned-moon phase. The air temperature does not influence the pattern. Even if it is cold, the birds will fly at night provided there is sufficient lunar illumination. But if it is cold and the level of lunar illumination is low, the birds become very torpid, reaching skin temperatures as low as 10°C (50°F). These data illustrate beautifully how flexible the abandonment of endothermy can be under different environmental circumstances. Indeed, there is no reason not to assume that this kind of facultative use of torpor in response to food availability and environmental conditions did not occur in other Late Cretaceous birds as well. As I have suggested regarding small archaic mammals, why would a small bird stay endothermic if there was no fitness benefit to being warm?—unless, of course, the bird was breeding.

Whether or not one would describe the African nightjars as hibernators is, in some respects, a pedantic question. Any animal that has the daily capacity to cope with a body temperature of 10°C (50°F) for several hours has the genetic and biochemical toolkit to avoid death through ischemic damage—that is, damage caused by a serious restriction of oxygen supply to the tissues. All hibernators and daily heterotherms have this capacity, but nonhibernators do not.

Although hibernation is rare in birds, it is not rare in mammals. Initially I did not know that my research on hibernation and torpor in the 1990s and thereafter would help me formulate ideas about what Schrëwdinger might have been like and how it fit into the story of the evolution of endothermy. The Holarctic Hibernation Paradigm muddied my early thinking. Holarctic biologists, in their highly seasonal high-latitude hometowns in Europe and North America, had generated the vast majority of data, conclusions, and opinions on hibernation. The revelation to me and other "Gondwanan" biologists, such as Gordon Grigg, Stewart Nicol, and Fritz Geiser, that hibernation and daily torpor did not evolve in temperate climates got us thinking about the evolution of endothermy from a different perspective. We now argue that endothermy and its abandonment—that is, hibernation and daily torpor—are two sides of the same coin. Endothermy and hibernation evolved during hot, tropical nights. To understand how the ancestral monotreme, marsupial, and placental mammal survived the Chicxulub event, we need to know how capable they were of using long-term hibernation.

The echidnas are excellent hibernators, but their closest relatives, the platypuses, are not, or at least not that we know about. Logic would therefore suggest that the ancestor of the platypuses and the echidnas could hibernate but that the capacity was lost in platypuses after the monotreme lineage split into platypuses and echidnas. This line of reasoning holds, of course, if the split occurred after the Cretaceous-Cenozoic boundary. But there is quite a bun-fight among monotreme scholars about when the split occurred, and the divergence of opinion is huge. Fortunately, entertaining as it is, it does not affect our story.

Stewart Nicol from the University of Tasmania has studied short-beaked echidnas (*Tachyglossus aculeatus*) throughout his career. He was one of the first evolutionary physiologists to obtain long-term and continuous body temperature measurements from a free-ranging mammal (Figure 10.6). His data have never failed to cause a stir at any of the hibernation conferences that I have attended. Echidnas are amazing hibernators, and some of their hibernation characteristics are truly unique. The males, for example, are distinctly rude but evolutionarily smart. They arouse from hibernation in midwinter, seek out the hibernating females, stir them into a semi-awake state, "and then mate with them while their body temperatures are still low." "Cool sex," Stewart calls it.[15]

There is no reason whatsoever for an echidna to maintain a constant elevated body temperature throughout a Tasmanian winter. Why? It is simply a waste of energy, which is difficult to find in Tasmania in the winter. Also, by avoiding activity, they can avoid being run over by Tasmanian bus drivers. Indeed, even in the tropics of Queensland, echidnas hibernate: tropical hibernation. They do not hibernate for as long as the Tasmanian echidnas do, but they hibernate nevertheless.

What is important to an echidna, and to all creatures on earth, is reproducing. They need only be warm enough for as long as it takes to produce offspring and see them off into the world. In the previous chapter I discussed Gordon Grigg's remarkable data showing how well echidnas regulated their body temperatures when they were incubating their eggs. The same strict

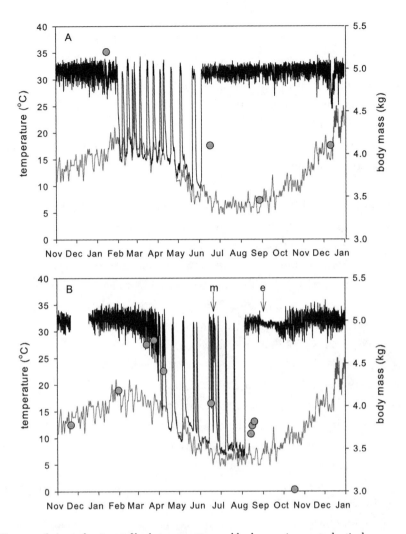

Figure 10.6. Annual pattern of body temperature and body mass in a reproductively active male (A) and female (B) echidna in Tasmania in the same year. Black line: body temperature; gray line: soil temperature measured at 20 cm (7.9 in); circles: body mass (right-side *y*-axis). The male entered hibernation in summer (1 February) after building up fat reserves in spring and early summer. The female reached maximum mass and entered hibernation much later (4 April). Unlike most other hibernators, echidnas may move to another location during interbout arousal periods. The male ended hibernation in early winter (4 June) and was found mating with the female on 17 June (m). The pregnant female then reentered hibernation, and her final arousal from hibernation was on 5 August. Shortly after this she entered a nursery burrow and laid an egg (e). Incubation of an echidna egg takes 10–11 days, during which the body temperature remains very stable, and in Tasmania the female typically stays in the burrow with the young for 23–48 days before leaving it in a plugged burrow while she forages. When she first emerges from the nursery burrow, her body mass is at its lowest. Males reach their minimum mass at the end of the mating period. (Data used with the kind permission of Stewart Nicol, University of Tasmania; redrawn from Nicol, S. C., "Energy Homeostasis in Monotremes," *Frontiers in Neuroscience* 11 [2017]: 195)

control of body temperature can be seen in Stewart Nicol's data for echidnas, shown by the arrows following arousals in the females. These data provide compelling support for the Farmer's Parental Care Model, which I discuss in the final chapter.

I have no doubt that the capacity for echidnas to hibernate as we see it today is the condition that the ancestral monotremes possessed at the Cretaceous-Cenozoic boundary, which enabled them to survive Chicxulub.

Understanding the origins of marsupial hibernation is simpler because in the family tree of the modern marsupials the possums branched off first, and they are hibernators. The marsupial that colonized Australia from Antarctica around 50 million years ago was probably also possum-like. There are two marsupial families that have hibernators today, the Burramyidae, the pygmy possums, and the Acrobatidae, the feathertail gliders. All are tiny marsupials, less than 20 g (0.71 oz), and are insectivorous. The glider also includes nectar and pollen in its diet.

Fritz Geiser is a German-born "Australian." He is also the world expert on marsupial hibernation and torpor. Fritz works at the University of New England, Armidale, on the Tablelands of New South Wales. His earliest work was on two species of *Cercartetus* pygmy possums.[16] Their body temperatures decreased to around 5°C (41°F) when in deep hibernation. These tiny marsupials spent up to a week hibernating before they aroused for a short period and then reentered hibernation. But Fritz and his longtime research collaborator Gerhard Körtner have shown that the master marsupial hibernator is the mountain pygmy possum, *Burramys parvus*, from the Australian Alps (Figure 10.7). They have recorded it as hibernating for more than a year, nonstop, in the laboratory.[17]

In collaboration with Thomas Ruf from the Veterinary University in Vienna, Fritz has published the most recent review of torpor and hibernation in birds and mammals.[18] Their data show that there are a total of six marsupials that hibernate: four pygmy possums, *Acrobates pygmaeus*, and *Dromiciops gliroides*. So clearly hibernation has been retained in the marsupial family tree right up to one of the most recent branches at which the Acrobatidae split off from the other Australian possums. In other marsupials

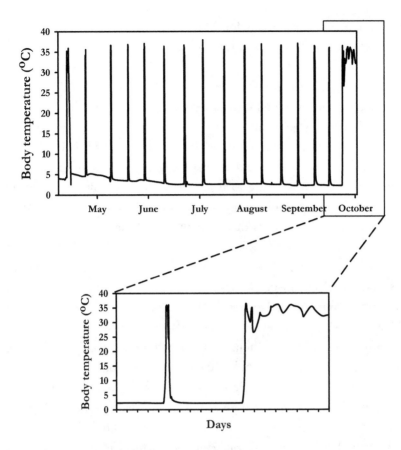

Figure 10.7. Body temperature data for the mountain pygmy possum, *Burramys parvus*, the best-known marsupial hibernator, obtained by Gerhard Körtner from free-ranging animals adjacent to the Charlotte Pass Village in Kosciusko National Park, New South Wales. The extracted graph shows the animal's final interbout arousal, followed by complete arousal to normal body temperature a few days later to end the hibernation season. (Data kindly provided by Gerhard Körtner and replotted; original data published in Körtner, G., and Geiser, F., "Ecology of Natural Hibernation in the Marsupial Mountain Pygmy-Possum [*Burramys parvus*]," *Oecologia* 113 [1998]: 170–178)

the capacity for true hibernation has been lost. However, there are twenty-six species of marsupials that also routinely use daily torpor.

The Australian marsupials inherited their capacity for hibernation from their closest South American relatives, the Microbiotheria, which includes the Monito del monte, Spanish for "little monkey of the mountain"

(*Dromiciops gliroides*). Today the Monito del monte lives in the highland forests of Chile and Argentina. It is an excellent hibernator.

It would be easy to assume that hibernation was an ancient ancestral condition in mammals, based on what is seen in marsupials and echidnas today. Especially in marsupials, it is in the species in the clades that branched off at the base of the family tree where hibernation and daily torpor occur most frequently in modern species today. But we would not reach the same conclusion if we were to have based our observation on placental mammals. That is because, until fairly recently, hibernation was unknown in any species in the clades that branched off at the root of the placental mammal tree.

For a long time it has been known that bears, marmots, and ground squirrels—North American or European species—hibernate routinely during winter. Indeed it is from data obtained from these species that the Holarctic Hibernation Paradigm was constructed. The clades that branched off at the root of the placental tree, though, are the Afrotheria (elephants, dugongs, aardvarks, hyraxes, golden moles, elephant shrews, and tenrecs) and the South American Xenarthra (sloths, anteaters, and armadillos). If hibernation could be found in these placental mammals, a much stronger case could be built for hibernation being the ancient condition.

Mariella Superina and Patrice Boily, from the University of New Orleans and Western Connecticut State University, respectively, were the first to find convincing evidence of hibernation in xenarthrans in the pichi, *Zaedyus pichiy*, a small desert-dwelling armadillo (Figure 10.8).[19] The average time spent in each hibernation bout was 75 hours. I must admit that I was quite relieved when these data were published, because I was starting to doubt whether the xenarthrans would ever come to my party and support the idea of ancient hibernation. There has been comparatively little research on South American mammals, so I would not be surprised if evidence of hibernation were found in other species as well.

I once took great pride in calling myself a "scholar" of the Afrotheria because these mammals are truly African in origin: mammals from my home

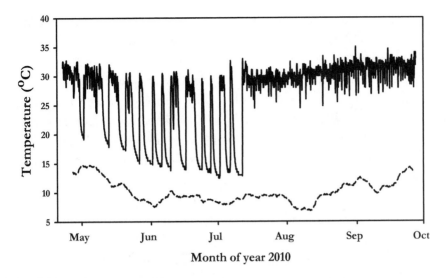

Figure 10.8. The first ever report of hibernation in the Xenarthra, the South American indigenous mammals that include armadillos, sloths, and anteaters, as measured in the pichi *Zaedyus pichiy*, from animals that were housed in outdoor enclosures. The upper trace is the pichi's body temperature, whereas the lower dashed trace is the burrow temperature. (Data kindly provided by Mariella Superina and Patrice Boily and replotted; originally published in Superina, M., and Boily, P., "Hibernation and Daily Torpor in an Armadillo, the Pichi [*Zaedyus pichiy*]," *Comparative Biochemistry and Physiology* A 148 [2007]: 893–898)

continent. But then the Schrëwdinger study came along and stuffed it all up, well and truly. The finding that the afrotherian ancestor was South American and not African came as a shock to me, yet it was strangely cathartic as well. It forced me out of denial about how I perceived certain fossils that had been claimed to be afrotherian but that were in the wrong place and I had conveniently chosen to ignore for a decade. These fossils did not fit the model of an African fauna isolated from the world since the split between Africa with South America about 100 million years ago. The maverick fossil is that of an elephant shrew—an elephant shrew not from Africa but from *North America*, the oldest elephant shrew ever described (Late Paleocene).[20]

"Splendid isolation" is a phrase immortalized by the famous American paleontologist George Gaylord Simpson in his description of the isolation of the South American fauna. In Africa, though, 100 million years of splen-did isolation of mammals did not occur—the Afrotheria did not have an African origin.

The O'Leary study found that two South American fossils were *part* of the afrotherian family tree![4] They were ungulates, hoofed mammals, bulky creatures that walked and foraged much as do modern peccaries. One was named after the famous twentieth-century biologist Thomas Hux-ley. The *Thomashuxleya* fossil is about 54 million years old. These extinct South American mammals are sandwiched between the lineage leading to the elephants and that leading to the hyraxes. The ancestral elephant must have been South American, and it must have looked something like *Thomashuxleya*.

What a shocker this was for scholars of the Afrotheria. The best explana-tion for the origins of the living Afrotheria pointed directly at South Amer-ica. All of the modern afrotherian lineages, that is, those leading to the el-ephants, hyraxes, dugongs, aardvarks, elephant shrews, tenrecs, and golden moles had ancestors that made their way to Africa from South America via North America and Europe. On their way an elephant shrew, *Apheliscus*, died and was fossilized in the Bighorn Basin in Wyoming.[20] The term "Afro-theria" became a misnomer overnight—devastating to Africans, worse than robbing Americans of baseball and apple pie.

For the moment, however, let's continue to call them the Afrotheria, be-cause as yet there is no new name. If we are to find hibernation in the Afro-theria, the most likely candidates need to be the right size, about the size of an Alpine marmot, let's say less than 5 kg (11 lb) or smaller. Do not turn to the size of bears for reference; they are really huge outliers in hibernation terms—completely atypical, enormous hibernators. Using the body-size line of reasoning we can immediately discount the elephant, the dugong, and the aardvark. They are too big.

Hibernation has been investigated in hyraxes and has not been found. This is, perhaps, not surprising because the three genera of hyraxes that ex-ist today are miniaturized, grazing ghosts of Africa. Modern carnivores and ungulates from North America invaded Africa via Europe and Asia about twenty million years ago. When these advanced placental mammals en-

tered Africa they found enormous herds of hyraxes grazing on the grasses of
the savannahs. These were not the small 2- to 3-kg (4.4- to 6.6-lb) hyraxes of
today. Some were as large as a small rhino. There were multitudes of hyrax
genera. They were the dominant grazing mammals in Africa by far.

Sadly, the hyraxes provided no opposition to the fast, modern placental
carnivores, which were, by this time, fully digitigrade. The hyraxes were
not fast because they still retained the flat, plantigrade feet of the ancestral
mammal. There was nothing fast that could chase them before the modern
carnivores arrived, so there was never any selection for the longer limbs and
reduced numbers of toes that occurred in ungulates. There were creodont
carnivores in Africa, but these were not fast animals, either. The hyraxes
were also no match for the placental ungulates, the antelopes, bovids, and
zebras. They could not compete with the highly derived four-chambered
stomach of the artiodactyls or with the ability of zebras to flourish by eating
grasses of poor quality.

Within a very short period of time, the once great diversity of African
hyraxes was reduced to what it is today, a mere three genera, which lurk in
trees and on the most inaccessible rocky cliffs. Hyraxes most certainly do
not venture far from cover anymore. Hibernation would have been "weeded
out" of the hyraxes of twenty million years ago by virtue of their large body
sizes at that time. It may be the same phenomenon that occurred in Cenozoic birds that I discussed earlier. Moreover, hibernation and the fermentation of grasses are two completely incompatible physiologies anyway.

Our search for hibernation was now limited to the insectivorous afrotherians, the elephant shrews, tenrecs, and golden moles. Here we hit gold.

ELEVEN

Clown Mouse and Golden Mole

One day my colleague Mike Lawes burst into my office with the most extraordinary prediction: "Elephant shrews must be hibernators!" he blurted out, eyes wild under his top-heavy blonde eyebrows.

Elephant shrews are delightful small mammals that don't bite. They have enormous, endearing eyes and a highly mobile long nose that twitches, arcs, and scents the world. In the Afrikaans language they are called *Klaas-neusmuis*, which means "clown mouse." But they are neither mice nor shrews—they are related to elephants; they are afrotherians.

Elephant shrews have caused enormous ructions in mammal classification circles over the past few decades. Initially they were placed within the order Insectivora, which included insectivores such as hedgehogs, moonrats, gymnures, moles (the European kind), and shrews. But the advent of DNA sequencing changed everything. Elephant shrew DNA was not most similar to that of other Insectivora at all. It was most similar to that of elephants, aardvarks, golden moles, tenrecs, hyraxes, and dugongs. It was most similar to the mammals of the Afrotheria. This spelled the end of Insectivora and the solid establishment of Afrotheria at that time. The remnants of Insectivora, sans golden moles and elephant shrews, are called Lipotyphla now.

"Mike," I said, "calm down. What the hell are you talking about?"

I explained to him that I thought elephant shrews would be the *last* mammals on the planet that I'd expect to hibernate.

Together with his postgraduate student Lizanne Roxburgh, Mike had been working on the current hot topic of the time, risk-sensitive foraging. The theory argued that there are two extremes of foraging that evolved in animals: risk-prone foraging and risk-averse foraging. In risk-prone foraging, animals take risky options that can result in big food windfalls—or nothing. Mike had studied elephant shrews in the laboratory and come to the conclusion that they showed all the signs of risk-prone foraging.

"Okay, get me some animals and I'll investigate," I suggested.

The animals, round-eared elephant shrews, *Macroscelides proboscideus* (Figure 11.1), duly arrived, and I surgically implanted them with tiny temperature-sensitive radio transmitters. I measured the intervals between radio signals and from the resulting data could calculate the internal body

Figure 11.1. The elephant shrew *Macroscelides proboscideus*, a small insectivorous member of the Afrotheria that uses daily torpor extensively. This photograph was taken in southern Namibia in the wild. (Image by and used with the kind permission of Galen Rathbun, California Academy of Sciences, San Francisco)

temperature of the animals. Some animals were given as much food as they could eat, and others were put on a mildly food-restricted diet.

Mike was right, sort of. The animals did not hibernate, but they certainly used daily torpor, especially those on the diet.

I was surprised, and from that moment onward my blinkered expectations about which animal might and which might not use torpor and hibernation were permanently removed. As I realized many years later, my thinking had been muddied by the dominating ideas of northern hemisphere biologists who were nurturing the Holarctic Hibernation Paradigm. These were the first data ever to be obtained for elephant shrews except for those found in a manuscript published in French reporting torpor in *Elephantulus rozeti* from Morocco.[1] The data set the precedent for much subsequent work on torpor and hibernation in elephant shrews. This was the first confirmation that a capacity for torpor resided within the afrotherian family tree, because I did this work before my Madagascan endeavors with tenrecs. Mike, Lizanne, and I published the data arguing that elephant shrews showed evidence of an ancestral capacity for torpor in placental mammals.[2]

As much as I was delighted about our results, I was also a bit confused. Why did these animals not have big bushy tails that they could wrap around themselves to minimize heat loss from their bodies during torpor, as dormice do? I investigated these odd creatures further and established that they never build themselves a nest, as do rats and mice and many other small mammals. In other words, they do not seem to care about conserving heat. It was at this point that I realized that heat conservation is exactly what an animal does not want to achieve if it is energetically beneficial to slide into a cold hypothermic state on a daily basis. Again I was trying to free myself from the Holarctic Hibernation Paradigm, which seemed to have become permanently etched into my neocortex.

The same concept would apply during heating. If a small mammal were to benefit from the outside air temperature, as ectotherms do, an insulating nest would prevent this from happening. Was it possible that certain small mammals were actually using the heat of the environment to warm up from torpor instead of using their own body heat generated by their metabolism? This was an important question for me to answer, because if the answer was yes, it was possible that this might reflect an ancient capacity, one used by the very first endotherms, to toggle between ectothermy and endothermy.

At the time I did not have the specialized equipment necessary to test the idea, so I went to work with Fritz Geiser during a three-month-long sabbatical in Armidale, Australia, in 1997. We worked on a small marsupial called a stripe-faced dunnart, *Sminthopsis macroura* (Figure 11.2). It is mouse sized and reaches a maximum mass of about 25 g (0.88 oz). It, too, has a long pointy nose. It is found throughout the northern and central desert regions of Australia. We wanted to test the idea that small mammals in the semitropics can arouse themselves from a torpid state by using the temperature of the air. Until that time, all we knew about the costs of arousing from daily torpor was what we had learned from the laboratories in the northern hemisphere, and their protocols were always the same. Their procedure was to keep an animal at a constant temperature, let's say 15°C (59°F), and continuously measure its body temperature and oxygen consumption,

Figure 11.2. The stripe-faced dunnart, *Sminthopsis macroura*, a small insectivorous marsupial in which daily torpor has been well studied. (Photo by and used with the kind permission of Gerhard Körtner, University of New England, Australia)

which are measures of their metabolic rate. The animals were also kept in a light regime similar to that of a northern hemisphere winter, something like 16 hours of dark and 8 hours of light. Light is a remarkable synchronizer of an endotherm's daily internal cycles, such as metabolism and body temperature. All mammals and birds, even blind ones, are hard-wired to the earth's light-dark cycle.

A popular small mammal used in daily torpor studies in Europe at the time was the Djungarian hamster, *Phodopus sungorus*. This delightful creature occurs naturally on the Siberian steppes. All northern hemisphere studies on small mammals, not only those on *Phodopus*, showed essentially the same thing. At the onset of darkness, the animals would start to produce internal heat and rewarm from their torpor body temperature, let's say 16°C (61°F). Rewarming would take about one or two hours, but the cost involved during this relatively short arousal time would be enormous. The heat was produced through non-shivering thermogenesis in the animal's brown fat deposits (described in Appendix 1). Yet there was still an energetic profit to be made by entering torpor daily, for a nocturnal mammal commencing in the early hours of the morning.

Although I did not know it at the time, what I was really trying to understand was the kind of thermal fluxes that Schrëwdinger might have faced at the end of the Cretaceous. In the hot, steamy days of the Paleocene, a constant temperature of 15°C (59°F) day and night, the protocol used in the laboratory, simply did not occur. In Europe during the Early Paleogene, the daytime temperatures in Celsius would have been in the low 30s (~86°F), and the nighttime temperatures might have decreased to the low 20s (~68°F). If the first Cenozoic mammals had used daily torpor, they would not have incurred anywhere near the costs of arousal that the modern *Phodopus* estimates were giving. And anyway, why would the ancestral mammals have needed to use their own heat to rewarm every day when they could use the heat of the air? This was the question at hand. Rephrased, is it possible that some animals can rely on daily cycles of air temperature to cool and heat them in and out of torpor every day at a fraction of the cost of rewarming from a flat 15°C (59°F)? Thermal hitchhikers is what these mammals would be, and if they showed close synchronization with the air temperatures, their patterns would not be much different from those of ectotherms. In some modern mammals these patterns may not be much different from those that occurred in the earliest mammals that were experimenting with endothermy, such as dogtooth.

Fritz's colleague, Gerhard Körtner, my old friend from my Marburg days in Gerhard Heldmaier's laboratory and an electronics whiz, helped me to design a temperature cycle that was a close approximation to what the dunnarts would probably experience in the wild. We also synchronized the temperature cycle to the light-dark cycle. Then we measured the dunnart's metabolic responses to these cycles.

The dunnarts entered torpor a few hours after midnight, when the air temperature had almost reached a minimum.[3] They turned down their metabolism, their internal heat production, by 80 percent. This is a good time to enter torpor, because the difference in temperature between the dunnart's body and that of the air immediately prior to its entry into torpor is the greatest of that at any time of a full-day cycle. Consequently, heat quickly leached out of the dunnarts, and their body temperatures plummeted toward that of the low air temperature. When the air temperature started to increase again during the morning, which would, of course, occur in the wild at sunrise, the dunnarts warmed up. They hitched a thermal ride on the air temperature, and they did it for free; they did not use their own heat production to heat themselves up until they had almost reached their normal body temperature.

These data showed that arousal from torpor need not be as expensive as the *Phodopus* estimates from the laboratory were showing. If a small mammal lives in a region of the world where there are large changes in air temperature between the day and the night during winter or summer, it can theoretically hitch a ride on the sunrise heat and save an enormous amount of energy. Fritz and his colleagues later showed that marsupials in the wild were using external heat much more routinely and regularly than we had ever imagined.[4]

I needed to confirm whether this was really happening in the wild in placental mammals, especially in the afrotherians. Was the ancestral placental mammal any different than the ancestral marsupial and monotreme at the end of the Cretaceous? I went back to South Africa fired up. My research candidate was obvious: the elephant shrew. My postgraduate whom I wanted to conduct the work was not.

Nomakwezi ("the morning star") Mzilikazi is a Xhosa girl who came into my life and tore me out of a complacent, privileged, white preserve. She made me desperately aware of my past, my upbringing, my limitations, and my fears. When we met, Kwezi was a just-fledged BSc honors student, the first ever black African postgraduate student I had supervised.

After she had completed her MSc with me, Kwezi took on the task of helping me to understand how placental mammals use daily torpor in the wild. For her main study we chose the rock elephant shrew, *Elephantulus myurus*. These elephant shrews do not look much different than *Macroscelides*, but they are a bit bigger.

We conducted our study in the Weenen Game Reserve in KwaZulu-Natal. Weenen is a small reserve with typical African savannah vegetation: a fifty-fifty mix of grasses and trees, mostly thorny acacias, and snakes, spitting cobras—*imfezi* in Zulu—lots of them. Kwezi was petrified of *imfezi*, which squirt venom at your eyes. On our first trapping trips she insisted that I walk in front of her, and for good reason. I got spat at on my first outing. It was a miss, because I turned my face to the side just in time and took it on the cheek. Kwezi finally managed to persuade the park management to give her a ranger, Baba Dladla, to walk in front and take the hits; I could not be with her all the time.

Baba Dladla could not figure her out. What was a young Xhosa woman doing in the middle of Zulu country catching elephant shrews? Xhosas come from the Eastern Cape of South Africa, where Nelson Mandela grew up. Mandela was a Xhosa. It was not only I who was mesmerized by this young black female biologist, a truly "New South Africa" thing, a very brave soul. As odd as it may sound, there was simply no precedent for young African women studying evolutionary physiology. It was wonderful watching the trail of bafflement that Kwezi left in her wake, wherever she went. It was wonderful seeing a young black South African taking the new opportunities available with wide-open arms, without demands or entitlements, setting a precedent. She seemed to bear little apparent resentment, and in this regard she was remarkably fortunate, because she could focus on her interests in life.

The plan was to capture elephant shrews, surgically implant them with small temperature loggers, and then release them again. After three months we planned to recapture them and retrieve the loggers. The plan was to repeat this four times, which would give us a whole year of data. Simple, we thought.

At the start, in winter, we caught eighteen animals and recaptured thirteen of them three months later. This was going to be a cinch. The data for the first three months were exciting. The animals were using torpor virtually every day. Some allowed their body temperatures to decrease to as low as 7°C (45°F) when in deep torpor, but, for the most part, the torpor body temperature was about 15°C (59°F), around 4°C (7°F) higher than the air temperature. They were synchronizing their use of torpor with the daily cycles of air temperature, which were impressive for winter. Black body temperatures ranged from a minimum of about 10°C (50°F) at 6 a.m. to a high of nearly 40°C (104°F) around midday. (A perfect black body absorbs all incident radiation and reflects nothing. Animals behave like black bodies. Copper model animals painted matte black were used to measure black body temperatures in the shade in the field.) The elephant shrews hitched a perfect ride on the sunrise heat and reached their normal body temperatures after about three hours.

In spring we caught another eighteen animals, implanted them, and let them go. Three months later we returned to recapture them.

The site was unrecognizable. A fire had ripped through the park, as fires routinely do in savannahs. The only things left standing on the windswept, charcoaled landscape were the acacias, which survive fires easily once they reach a certain size. In three weeks we caught only two animals, but two very valuable, scorched, spring-data animals. They showed the same pattern as the winter animals.

Elephant shrews were doing more than just going into torpor every day. They were adjusting their activity patterns to suit their need to enter into torpor.[5] Normally elephant shrews are active in the early evening and the early morning, in what is called crepuscular activity. They follow this pattern with some predictability in summer, when they breed. But in winter they modify their activity time to the afternoon and early evening only. They shift their activity pattern to synchronize with the temperature cycle of the air so they can enter torpor and arouse using ambient sunrise heat.

The data confirmed that placental mammals, like the several species of small marsupials on which Fritz Geiser and Gerhard Körtner had so painstakingly obtained data in the field, have the capacity to synchronize their heating and cooling patterns with the sun's daily cycles.

I do not regard this capacity as something that is new or derived. In other words, I don't consider it an adaptation that evolved during the Cenozoic. I'd argue strongly that it is an ancient ancestral condition that existed in the earliest placental mammal, Schrëwdinger, and its ancestors, like dogtooth. Again, like Ben Smit's nightjar data, these data illustrated the flexibility and adaptability of the abandonment of endothermy when it is not required for reproduction and how it can be employed to survive the paucity of food following events such as wildfires.

The next team to study elephant shrews was headed by Justin Boyles, an American postdoctoral fellow collaborating with Andrew McKechnie, Ben Smit, and Catherine Sole from the University of Pretoria. On the hunch that hibernation might be found in elephant shrews that live in very arid regions, Justin Boyles headed for Namaqualand with Ben Smit to the farm Noheep, where Ben had done the nightjar work. They decided to test their implantable temperature-logging devices by swallowing one each, washed down with a bottle of whiskey. I'll leave it at that.

There are two species of elephant shrew that live in the mountains of Namaqualand together, sympatrically: the western rock elephant shrew, *Elephantulus rupestris*, and the Cape rock elephant shrew, *Elephantulus edwardii*. The Boyles team followed the same procedure that Kwezi had followed in Weenen; they caught animals, implanted them with temperature-measuring devices, released them, and recaptured them one to two months later.

Although both species used daily torpor, there was no hibernation—no torpor that lasted for more than 24 hours.[6] Only male Cape rock elephant shrews were recaptured, and they rarely used daily torpor. *Elephantulus rupestris*, on the other hand, regularly used daily torpor, with minimum body temperatures of about 8°C (46°F) recorded in some individuals. The difference in the expression of torpor by the two species living side by side was curious and might have something to do with Mike Lawes's predictions about risk-sensitive behavior. Is it possible, for example, that *Elephantulus rupestris* is a risk-prone forager and hence needs the capacity for daily torpor, as does *Macroscelides proboscideus*, which we discussed earlier, whereas *Elephantulus edwardii* may be the opposite, a risk-averse forager? And if this difference in foraging modes occurs, can it explain how two seemingly identical elephant shrews can live side by side with each other? We need tons more data to answer such questions.

Then Fritz Geiser got in on the act, in collaboration with Kwezi, who by now held a lecturing position at Nelson Mandela University in Port Elizabeth. They investigated potential hibernation in *Elephantulus edwardii* in the laboratory, where they could control the air temperatures and the amount of food available to the animals. Their data showed very convincingly that both males and females used daily torpor routinely and that in some instances torpor bouts lasted for nearly two days (44 hours): "Torpor bouts of elephant-shrews are intermediate in duration to those of daily heterotherms and hibernating mammals, but their body temperatures (T_bs) and metabolic rates are very low and similar to those of hibernating mammals."[7]

These were tantalizing data because they suggested that the distinction between hibernation and daily torpor might not be as rigid as we had always thought.

Fritz and another old colleague and friend from my Marburg days in Heldmaier's laboratory, Thomas Ruf, decided to resolve the distinction between hibernation and daily torpor once and for all. They assembled the biggest database ever on daily torpor and hibernation in birds and mammals.[8] Their analyses were conclusive: there is a very clear distinction between the characteristics of daily torpor and hibernation. Daily heterotherms are smaller, on average 33.6 g (1.19 oz), and occur at lower latitudes (on average, at 25° north and south). Hibernators are bigger, on average 2.4 kg (5.3 lb) and occur at higher latitudes (on average, at 35° north and south). But the characteristic that defined the two types of heterothermy best was the maximum bout length, that is, the maximum time that each species spent in a torpid state. In daily heterotherms, the maximum torpor bout length was, on average, 11.2 hours, but in hibernators it was a whopping 266.6 hours—about 11 days. There was also a huge difference in the minimum body temperatures attained during the torpid state. In daily heterotherms it was 16.9°C (62.4°F), but in hibernators it was a mere 3.9°C (39°F).

Based on these data, we must regard elephant shrews as the most "extreme" daily heterotherms. Within daily heterotherms they seem to attain the lowest body temperatures, and their maximum bout lengths sometimes, but rarely, exceed 24 hours. But they are not hibernators. Fritz and Thomas reached this conclusion: "Arguably, the primary physiological difference between daily torpor and hibernation, which leads to a variety of derived further distinct characteristics, is the temporal control of entry into and arousal from torpor, which is governed by the circadian clock in daily heterotherms, but apparently not in hibernators."

My interest in the Afrotheria provided the incentive to learn more about another small insectivorous afrotherian group, the golden moles (Figure 11.3). It has been known for some time that golden moles, when resting, show a daily pattern of body temperature that virtually tracks daily amplitudes of soil temperature, which indicates the capacity for daily torpor.[9,10] I wanted to know whether they could hibernate, but I could never convince a research student to undertake the daunting task of studying them. Golden moles are completely blind and burrow using their noses as wedges. Gram for gram, they are enormously powerful, especially in the front legs. You need to hold one in your hand to experience this. Their front toes have been reduced to two spade-like digging tools. They have rather skinny little back legs. Virtually all species occur in southern Africa, from the driest sand dunes of the Namib Desert to the forests of the Eastern Cape. Golden moles have the most beautiful soft, velvety fur that glints with iridescent gold-to-purple hues in sunlight. They are small, mostly around 50 g (1.76 oz), except for the giant golden moles from the forests of the Eastern Cape, *Chrysospalax trevelyani*.

Figure 11.3. Grant's golden mole, *Erimitalpa granti*, which inhabits the sand dunes of the Namib Desert. (Photo by and used with the kind permission of Galen Rathbun, California Academy of Sciences, San Francisco)

Studying golden moles is not easy because they are hard to catch. They are every gardener's nightmare, and they love to burrow a few centimeters (0.8 in) belowground in a newly sown veggie or flower patch. They are elusive and do not readily fall for live traps.

Then one day I got an email from Mike Scantlebury, who is British and currently works at Belfast University. Mike had a project going on golden moles (*Amblysomus hottentotus*) in the foothills of the Drakensberg Mountains. He assured me that he had "tons of willing student help" from the University of Pretoria to catch golden moles. Mike was collaborating there with Nigel Bennett, Africa's crackerjack mole man. They were working on a golf course in a mutual agreement between the greenskeeper and themselves. I don't live or work far from the Drakensbergs, so Mike wanted to know if I'd be interested in collaborating with them by doing the surgery required to implant the animals with miniature data-logging devices. What a silly question.

Golden moles do not respond well to being held in captivity, so we developed a modus operandi that minimized captivity time. I have to admit,

Mike and his students worked supremely hard, day and night, checking their traps every hour or so. Once a mole had been caught, the game was on. I'd be called, at any hour of the day or night, and from then I had about an hour to get my surgery ready. I sent my students to the pet shop to get earthworms to feed the moles after surgery.

Over a period of several weeks, we caught and surgically implanted fourteen animals over two seasons and released them back into the burrows where they had been caught. We waited about four months and then launched the mission to recapture them. We caught two animals. That's all, only two miserable animals, even with a massive investment of student effort and enthusiasm. But two animals were better than no animals at all, and, best still, both implanted devices produced high-quality and revealing, albeit starkly different, data.[11]

One mole exhibited no deviation from a constant average temperature of 33.9°C (93°F). The animal maintained this temperature for the entire thirteen weeks of measurement, exhibiting pure basoendothermy, not uncommon in subterranean mammals.[12] If this had been the only animal that we had recaptured, we would undoubtedly have reached the conclusion that golden moles cannot hibernate. But the other animal exhibited bouts of true hibernation on four occasions, with the longest bout lasting five days (Figure 11.4).

Science is a funny thing. Strictly speaking, we had no right to publish these data. Our hypothesis had been simple: golden moles are hibernators. To test this hypothesis we needed to be 95 percent certain that this was the case; that's the rule in science. This meant that if we had recaptured all fourteen animals that we implanted, at least thirteen would have needed to show hibernation for our hypothesis to be valid. But we recaptured only two animals, and only one hibernated, yet we managed to publish the data anyway.

We got away with a hibernation sample size of one animal because of the immense scientific value and interest in the data, which was greatly appreciated by the journal editor. When you implant a device into an animal and it takes a temperature measurement every thirty minutes and stores the data onboard for later retrieval, you end up with eight thousand data points that

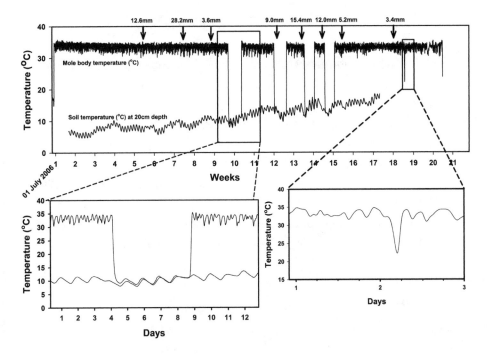

Figure 11.4. The soil temperature (lower line) and body temperature (upper line) of a free-ranging golden mole *Amblysomus hottentotus* measured over five months. Two sections of the body temperature trace have been exploded in the two lower graphs: a hibernation bout (lower left) and a torpor bout (lower right). The arrows at the top of the body temperature trace indicate rainfall events, with the numbers in millimeters giving the amount of rainfall. (Figure redrawn from Scantlebury, M., et al., "Hibernation and Non-Shivering Thermogenesis in the Hottentot Golden Mole [*Amblysomus hottentottus longiceps*]," *Journal of Comparative Physiology B* 178 [2008]: 887–897)

have recorded what happened thermally in that animal over four months. These are very valuable data obtained from a free-ranging, highly evasive subterranean mammal. They cannot be discarded!

I am not arguing that evolutionary physiologists should be excused from applying the hypothetico-deductive scientific method to test their ideas. But I would argue, strongly, that the data should be published because, when withheld from other biologists, they retard the process of understanding, in this case, the evolution of hibernation and endothermy in mammals. A mammal either hibernates or it does not. If it does hibernate, it must possess the suite of functional genes that encode for the biochemical processes that make hibernation possible. I'd argue at this point that we can make a sound, publishable induction about hibernation in golden moles, but we must continue to strive for a sound deduction to be made using the routine scientific process.

The golden mole data added to our dossier on the Afrotheria. This clade, which branched off first in the placental mammal tree, virtually at the Cretaceous-Cenozoic boundary, possesses functional genes for hibernation. Therefore, I deduced, it must have inherited these genes from its ancestor, from Schrëwdinger. Schrëwdinger must have been a hibernator. But I still had one more species of afrotherian to investigate to be absolutely sure. I needed to go to Madagascar to study tenrecs, and that is very hard to do if you cannot speak French or Malagasy.

TWELVE

Madness

.

The flight from Johannesburg to Antananarivo, or Tana, takes a little over three hours. From a window seat on the starboard side you get the first glimpse of the shoreline of the great red island anchored in the Mozambique Channel. It is hard to imagine that Madagascar and India were once part of Africa, somewhere near Sudan. The Madagascar-India proto-continent split off from Africa about 160 million years ago, drifted south for a while, and then headed east, into the Indian Ocean, opposite Mozambique. Then it got stuck, and, about 80 million years ago, India broke off and kept going, only this time it headed north, until it crashed into the European continent and threw up the Himalayas. Madagascar stayed put about 500 km (310 mi) east of Mozambique.

The flight path into Tana is picturesque. The mosaic of rice paddies, swamps, and clusters of bizarre little houses perched on islands in the rice paddies is a stark contrast to the red, unfruitful, eroded landscape between the coast and the city. It is vastly different than anything that you can see in Africa or anywhere else in the world. Madagascar is a smorgasbord of cultures blended from African genes in the south to Asian genes in the north, mixed throughout with French genes. And the people are small, except the French. They are also very poor, except the French. I have seen poverty in Africa, lots of it. Poor people in Africa always seem miserable. Poor people in Madagascar seem sort of indifferent. It is hard to make a living on an island, but the Malagasy have managed to do so, successfully—at the terrible expense of the greatest and most unique paradise of biodiversity that once existed on earth, as I explain further in the next chapter. And the Malagasy are skinny. In all my visits to Madagascar, I have rarely seen obesity.

Tana is Merina country, the "king-of-the-castle" spot in Madagascar. The Merina are the dominant tribe in Madagascar. They established the town Antananarivo, which means "the place of a thousand people," where it is, because it is cool and wet, perfect for growing rice and avoiding the terrifying heat of the lowlands.

The "taxi" ride into Tana is a rollercoaster into the Middle Ages, surrealism by the bucketful. The architecture shows many influences but has an indelible French finish. The buildings are small, mud-brick double-stories with wrap-around verandas fronted with delightful wrought iron grilles. In the countryside around Tana the Malagasy keep their livestock on the ground floor of their homes and sleep on the north side upstairs. Sleep location in buildings is serious stuff in many parts of Madagascar—*fady* stuff. *Fady* are more than just taboos. They are a set of time-honored cultural no-nos that glue the Malagasy cultures together. They are the rules of the ancestors, and they must not be broken. They are cultural adaptations, if you wish, to different regional challenges—rules that make things work on an overcrowded island. There are hundreds of *fady*, but *vasaha*—foreigners—are forgiven for not knowing them.

When I first went to Madagascar in August 2006, there were no supermarkets. In most towns in Madagascar there still are no supermarkets. Everything is sold on the street, with flies: meat, fruit, vegetables, rice in a hundred varieties, stacked in perfect cones, toiletries, clothes, pencils, pens, and girls—lots of very young girls. Prostitution is a colossal business in Madagascar, fueled mostly by old, gone-to-seed Frenchmen "visiting" from France. It is illegal, but ignored because it is vital to the economy throughout the country. The shagged-out dudes shopping or dining with their underage sex workers can be seen all over the island.

Fabien Génin, a French primatologist, joined me on my first trip to Madagascar. I met him at the airport in Johannesburg after his flight from Paris. He appeared through the green exit sporting a beret, worn for immediate identification, he said. Fabien knows southern Madagascar well. He was amused because within 48 hours of entering Madagascar I got the Aztec two-step: the worst diarrhea known to man. Fabien reckoned he knows how to beat it.

"When you first get to Tana," said he, "eat all ze foods you see, especially ze food covered wis ze flies—very nice. You'll get ze tummy problem once, but never again."

Right.

Fabien lived like a typical student—he never had money. He ordered the taxi driver to take us to his "darling" hotel—read cheapest hotel—in the center of Antananarivo, Hotel Lambert. Having never been to a whore-house in my life before, I checked in with the innocence and excitement of any tourist in a new city. Everyone was so nice and friendly, especially all the girls draped over the furniture and the assortment of old Frenchmen drinking coffee and eating croissants. The reality of where I was staying dawned on me later that night. I was kept awake the whole night by the incessant thumping on the walls of my room by the beds in the adjacent rooms.

With Fabien's immunity wisdom bubbling away inside me, we flew down to Fort Daufin, now known as Tôlagnaro, a very different town from Tana, on the southeastern tip of Madagascar. A hot, moist breeze blows off the Indian Ocean in Tôlagnaro, a pleasant relief from the smelly atmosphere of Tana and the Hotel Lambert's flesh market. This is Tandroy country. The Tandroy tribe has African origins. They are small, wiry, and fierce. The men carry spears with very sharp, pointy blades on the ends of narrow shafts of steel to ward off gone-to-seed Frenchmen from their girls and for other reasons.

Tôlagnaro is the gateway to Berenty, the only private reserve in Mada-gascar, owned by monsieur Jean de Heaulme. We headed there in search of tenrecs, the lesser hedgehog tenrec, *Echinops telfairi*, and the greater hedgehog tenrec, *Setifer setosus*. The de Heaulme family has conserved a unique mosaic of habitats in southern Madagascar along the banks of the Mandrake River. The surrounding countryside is spiny desert, but gallery forests of huge trees dominated by tamarinds (*Tamarindus indica*) grow along the banks of the river. The reserve drips with lemurs, especially the most charismatic of all, the ring-tailed lemur, *Lemur catta*. The de Heaulme family have been very accommodating to international researchers for decades, and there are many attractions to persuade scientists to undertake research in this place. The research facilities are free to researchers, and there is an exhausting diversity of intriguing plants and animals to study. But for us, at that time, when I still did not really know what I was doing, the most important bonus was that we did not need to hold a government research permit to work in a private reserve. I am not saying that this was necessarily a good thing, but just that it was very convenient for a quick scientific "look-see" before I committed to something bigger. The process of getting a research permit in Madagascar generates torment that can take years off your life.

My arrangement with Fabien was reciprocal. He was not really interested in tenrecs; he studies mouse lemurs in the spiny forest adjacent to the gallery forest. But he had no job and no money at the time, and he wanted to get back to Madagascar and his mouse lemurs. I paid for him to take a trip to Madagascar in exchange for his helping me with my preliminary investigations of tenrec endothermy and being my translator when I was speaking with Jean de Heaulme. De Heaulme does not speak English, and I do not speak French, except when ordering wine.

I really liked Berenty as a potential research site and promised myself that I would get back as soon as I could.

In April 2007 I was back in the reserve with Fabien. I got wise on my second trip and avoided Tana altogether. We found a flight that went to Tana from Johannesburg, via Tôlagnaro. Three hours, and we were there. But it took another six hours to drive less than 90 km (56 mi) to Berenty, so appalling are the "roads." Throughout Madagascar, the road infrastructure was allowed to deteriorate after independence, when the communists took over. Apart from a few new roads recently built by the European Union in the northwest, the roads, mostly, remain the same—completely worthless— one large pothole.

I was ready to make some measurements on hedgehog tenrecs, but we soon established that tenrecs are a delicacy and were hunted everywhere.

Hunting tenrecs is a top Tandroy sport. There was no tenrec *fady* in Berenty as in some parts of Madagascar, where it is forbidden to eat certain species. Also, there was a flood two weeks before our arrival that, as the Tandroy say, "washed all the tenrecs down the river."

We realized soon that our proposed research plan was not going to work. We had planned to catch animals, surgically implant them with tiny temperature-logging devices, fit them with collar-mounted radio transmitters, and let them go back into their natural habitat exactly where we had caught them. Easy. We would then locate them periodically by radio-tracking them and see what they were up to, and finally, after three months, catch them again and remove the data loggers. We were going to get the first glimpse ever into the free-ranging patterns of endothermy in one of the world's most curious mammals. This was exciting stuff.

But alas, there were no tenrecs.

Apart from periodically stabbing one another and gone-to-seed Frenchmen with their spears, the Tandroy also use them to find tenrecs. Random short, sharp jabs into tree holes or holes in the ground elicit a squeal if a tenrec is at home. Then the dogs and machetes take over. Everyone in Madagascar seems to carry the ubiquitous machetes.

At Berenty, all local Tandroy work for de Heaulme on his massive sisal estate. But the Tandroy could not find us any tenrecs in the spiny forest or the gallery forest.

Fabien hatched a plan, and started waving around some of my ariary, Malagasy currency, to make something happen. By morning we had six tenrecs, six lesser hedgehog tenrecs, *Echinops telfairi* (Figure 12.1). We asked the catchers where they had caught the animals. "In the sisal," they said. Our plan to release animals into a pristine environment to measure their endothermic responses within that environment was sunk. A sisal plantation is hardly a pristine environment.

Fabien hatched another plan. He took me to the ruins of a small old "zoo" in the gallery forest, not far from a tourist restaurant. There we found enclosures that at one time had housed crocodiles. The walls were about a meter high, and the enclosures were covered with elevated wire-net frames like those of an aviary, about three meters high. Nothing could get in, and

Figure 12.1. The lesser hedgehog tenrec *Echinops telfairi,* which is virtually ectothermic when not breeding. (Photo taken by the author at Berenty, southwestern Madagascar)

nothing could get out, except through the gate. Fabien successfully negotiated with de Heaulme to secure one enclosure for our tenrecs.

We were ready to do the surgery. We found a small wooden hut in the research village and scrubbed the table thoroughly with 70 percent alcohol. It was absurdly hot in the hut. Once my hands were sterile, Fabien had to wipe my forehead every few minutes to stop the sweat from falling onto the operating area. My shirt was quickly drenched. We had brought along with us a cheap Chinese stovetop pressure cooker that we bought in an Indian trading store in Tôlagnaro to sterilize our surgical equipment. It worked beautifully.

We implanted the tenrecs with the temperature data loggers and ensconced them comfortably in their new home. We hired a Tandroy and gave him the responsibility of looking after the tenrecs for three months. He caught Madagascan hissing cockroaches (*Gromphadorhina portentosa*) in the gallery forest every day and tossed them into the enclosure. He stopped feeding, he said, only when the numbers of roaches increased, which meant, he reckoned, that the tenrecs were "sleeping."

Three months later, we were back. All six tenrecs were alive and well—a great start. We excitedly got going with the surgery to remove the loggers first thing the next morning. The first one to come out went straight into the reader connected to the laptop—nothing. There were no data. The device had malfunctioned. There was a stony silence between us. The second one came out—nothing—gulp. The third one came out—nothing. I snapped, swore, and smoked. We must have botched something, somewhere. I was shaking with disappointment and desperate to find someone, or something, to blame.

In a very black mood, we walked the shortest and hottest route to the tourist restaurant for lunch, in silence. My guts were bubbling again, so I had beer for lunch, lots of it—Three Horses Beer, known locally as THB. Fabien ate everything he could order; "very nice," he said.

We tried to figure out what had gone wrong but could not find an explanation. I'd done the exact same procedures hundreds of times with other animals, and they had always worked. It was only months later that I figured it out. The new models of the stainless steel–covered devices that we had used were not waterproof like the earlier models, because new European Union rules had forbidden the use of solder in electronics. Body juices had seeped into the devices and destroyed the circuit boards.

We decided to walk back to the research village along the tourist path through the gallery forest. It is a delightful walk. There are plenty of comical troops of ringtail lemurs to see and bizarre sifakas that waltz and skip across the path—retarded poetry in motion. They calmed us down into a phase of acceptance, a numb reality. We decided to remove the last three devices and go home.

Device four—data! Device five—data! Device six—data! We marveled at the wretchedness of probabilities, and I danced around the hut. The data were more exciting than I could ever have imagined. Data from three animals are so much better than no data at all. Fortunately, the data showed exactly the same patterns in all three animals, which made it easier to persuade the editor of the *Journal of Comparative Physiology B* to publish them, even with the small sample size (Figure 12.2).[1]

Not once in three months did any of the tenrecs attempt to maintain a constant body temperature during the daytime. Their body temperatures simply tracked the air temperature. And not once, either, did they seem to maintain an obvious active-phase body temperature when, we presume, they became active in the early evening after being heated during the afternoon. The tenrecs were not keeping a constant body temperature during their active phase and entering torpor during their rest phase as did typical daily heterotherms. Instead they were being ectothermic during their rest phase and were producing heat proportional to their own activities during their active phase. Further, they were not producing a consistent amount of heat to defend a constant body temperature. They were not acting as mammals are supposed to act, physiologically speaking, if they were to be called true endotherms.

We made these measurements in the middle of the austral winter, so the animals were not breeding. They were maintaining life at the lowest possible metabolic level, waiting for summer to reproduce. They had no need to maintain a constant body temperature. In winter, lesser hedgehog tenrecs are virtually metabolically switched off. They spend their days wedged into tiny spaces in tree holes, and probably do only a limited amount of foraging at night. There is not much to eat in the spiny desert in winter.

I also realized that the spines on the hedgehog tenrecs might have less to do with direct defense than we might intuitively have imagined. Both the lesser and the greater hedgehog tenrecs roll up into spiny balls when threatened. But, as I point out in the next chapter, this does not seem to protect them from being eaten by predators such as snakes. What the spines *are* very effective at doing is wedging the rolled-up animals into their crevices, for example, into tree holes. Hedgehog tenrecs inflate their lungs when threatened in their resting places, which forces the spines into the surrounding woodwork. They are then virtually impossible to dislodge; they are anchored. For an animal that spends virtually every day in a state of torpor, the

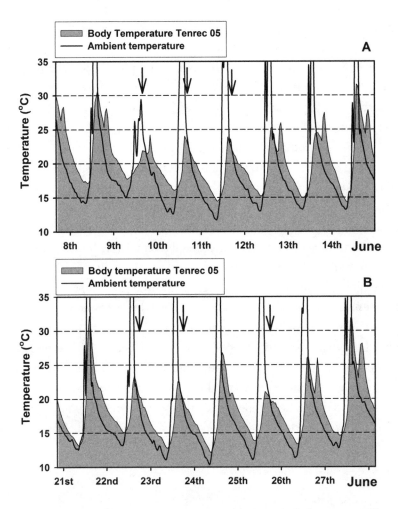

Figure 12.2. The daily patterns of body temperature of the lesser hedgehog tenrec, *Echinops telfairi*, measured over a week showing a three-day hibernation bout (indicated by arrows) during which the animal's body temperature did not exceed 25°C (77°F). (Figure redrawn from Lovegrove, B. G., and Génin, F., "Torpor and Hibernation in a Basal Placental Mammal, the Lesser Hedgehog Tenrec *Echinops telfairi*," *Journal of Comparative Physiology B* 178 [2008]: 691–698)

capacity to remain wedged into tight crevices and thus avoid predation must infer a huge fitness benefit.

As I flew home the same questions kept cropping up: What happens in summer? What happens when the tenrecs breed? What happens when they are pregnant or lactating? What do the males do? I started to question the concept of daily heterothermy. What if being cold, or hypothermic, was their normal condition, and being warm was the daily exception—were they daily endotherms? This is exactly what the earliest mammals, dogtooth and *Hadrocodium*, which showed the first characteristics of heat production for thermoregulatory purposes, might have done at night. More specifically, they might have been nightly endotherms, but not throughout the year.

The game was on. I had to have the answers. I was fired up and ready to commit all of my research resources to get them. But *Echinops telfairi* was not going to be the best model tenrec to study. *Echinops* is restricted to the southwestern deserts of Madagascar. I needed a tenrec that was more widespread across Madagascar so that I could compare tropical with semi-tropical populations. I also needed a site where I could do the research. I needed a site where the tenrecs were protected or, even better, where there was a *fady* against eating them. And I needed dedicated research students, at least two, who loved rice, hopefully could speak French, and were prepared to live and work in Madagascar for several years. To begin, I needed to make a recon trip into other parts of Madagascar.

THIRTEEN

Ankarafantsika

A nkarafantsika is a very special dry deciduous forest in western Madagascar. It is small but is home to a species of bird and two species of mammal that occur in that forest and nowhere else in the world. The endemic mammals include the big-footed mouse, *Macrotarsomys ingens,* and the Lac Ravelobe mouse lemur, *Microcebus ravelobensis.* The most spectacular endemic bird is Schlegel's asity, *Philepitta schlegeli.*

A lovely, scenic lake, Lake Ravelobe, brimming with sacred crocodiles, splits Ankarafantsika. It is from the lake that *Microcebus ravelobensis* gets its name. The lake is named after Ravelobe, a Malagasy patriot who participated in World War II. After the war, on his return to Ankarafantsika, he requested that the French colonists govern the area, that is, Ambatoboeny. After his request was denied, he dedicated his career to robbing people traveling the road through Ampijoroa that leads to and from the coast.

But the legend about the lake goes back to long before World War II. During the reign of Radama between 1810 and 1828, King Andriamisondrotramasinarivo—that was indeed his name—ruled the Ankarafantsika region. At the time, Radama and his Merina tribe were busy dominating all Malagasy tribes, but King Andriamisondrotramasinarivo was having none of it, so he fed himself and his family to the lake's crocodiles alive rather than submit to Radama.

The park's headquarters are situated adjacent to the town of Ampijoroa beside the lake. It has a small research tent camp, but researchers have access to other useful amenities provided for tourists, such as a restaurant and flush toilets, sans toilet paper or seats. Importantly, it has electricity, sometimes.

Ankarafantsika has two species of tenrec that were perfect models we could use to answer my endothermy questions: the greater hedgehog tenrec, *Setifer setosus*, and the common tenrec, *Tenrec ecaudatus*. *Setifer* is almost identical to *Echinops*, just a bit bigger and grayer. It also occurs throughout Madagascar, along with the common tenrec.

Unlike the eastern rainforests, the western forests have a long dry season, and most plants lose their leaves in winter. My first visit there, with a new MSc student, Kerileigh Lobban, was in the rainy season, and the forest was in full bloom. The trees and shrubs were in flower, and there were hoards of insects everywhere. It was the time of year when things really happen in Ankarafantsika, when animals and plants reproduce. We met a fantastic guide, Ndrema, who showed us around the tourist circuits because we were, at that stage, officially still tourists. On the first day we did a seven-hour hike through the forest; there were snakes, lemurs, snakes, snakes, and snakes—so very many snakes.

But we were there for tenrecs, so we persuaded Ndrema to accompany us on our first night walk. At that time, tourists were still allowed to walk in the forests at night. Keri was interviewed by *Live Science* on her first impressions of working in the forests at night. She said, "Nighttime in Madagascar forests is absolutely magical, teeming with weird and wonderful creatures— eyes glaring at you from all directions as you scan the forests with head-lamps, harmless boas slithering past ever so often, rustling of branches as the lemurs shift around in the trees . . . and the sounds—oh, the sounds! Eerie songs of the avahi and high-pitched squeaks of the mouse lemurs. Night-time in Madagascar is something to be experienced."[1]

Walking in a forest at night with a headlamp is like scuba diving at night. All you see is what the light reflects, about 10 percent of what is actually there. Of course, scuba diving is much more scary; you just know that there are things with big teeth watching you from behind, always. But in the forest at night it is the eyes that make the difference, the reflective eyes of nocturnal animals—lemurs, spiders, geckos, and moths—if they look your way. Huge hawk moths hovered in front of me, an arm's length away, probing the genitals of orchids and other plants whose names I did not know.

"Thwak, vrrrrr," I heard as a tiny scops owl swooped out of the dark and caught a moth in the beam of my headlamp. I felt a bit guilty as I watched the moth's wing scales flutter to the ground. The owl must have seen the moth in the beam.

Little birds sleep at night on the flimsiest of twigs overhanging the forest paths, twigs too dainty to bear the weight of sneaky predators—little orange kingfishers and robins, fast asleep, within grasping distance.

Ndrema was armed with a stick with a forked end. He dived into the forest on the side of the path and pounced on a tenrec, a big male. I got to see my first *Tenrec ecaudatus*. That night we caught another, unofficially, and saw several others. The next night we caught a greater hedgehog tenrec, *Setifer setosus*. There seemed to be plenty of tenrecs of both species at Ankarafantsika. I made a decision. Ankarafantsika was the site I had been looking for.

All I needed now was another research student.

As I said before, along with my former PhD student Andrew McKechnie, I organized the Thirteenth International Hibernation Symposium in Swakopmund, Namibia. A young Canadian woman, Danielle Levesque, approached me there and asked if she could join my team as a PhD student. Danielle was perfect for Ankarafantsika. She could speak fluent French, she had research experience (an MSc), she was experienced at doing implant surgery, and she had worked with a doyen of endothermy evolution, Gordon Grigg at the University of Queensland. I did not ask her whether she liked rice.

I had my team, Keri and Danielle, and I had my site, Ankarafantsika. It was time to get to work in Madagascar and get some more data!

After a long wait back in South Africa, Keri and I finally obtained the cherished research permit issued by the Malagasy National Parks to continue our search for the ancient characteristics of endothermy in Afrotheria. We were now researchers and no longer tourists. Danielle joined us from Canada. I scraped together the last of the research funds that I had been saving for the rainy day that had arrived. We bought and gathered together everything that we needed for the camp in Ankarafantsika—including a kitchen sink—and shipped it in a container to Antananarivo via the port of Toamasina.

Based on the shipping schedule and the estimated clearance time, we timed our trip to Antananarivo to meet the shipment and have it transported with us to Ankarafantsika, which is on the other side of the island.

But in Antananarivo we waited, and waited, and waited. The weeks went by, weeks of attempted bribes, arguing, screaming, shouting, and swearing. The customs office refused to release our equipment without a payment that was not required.

We decided to kill time and take a break from Antananarivo. We flew to Maroantsetra, the gateway to the Masoala Peninsula. I wanted to show my students my treasured hideaway, a tropical paradise that I'd visited the year before.

It was depressing. Everyone was running scared in Maroantsetra. Olivier Fournajoux, who lived with his wife and kids in Maroantsetra and ran the Ecolodge Chez Arol on the Masoala Peninsula, picked us up at the airstrip. Olivier was jittery. He told us not to go near the port area and to hide our cameras. Chinese-syndicated poachers were plundering Masoala. Their modus operandi was to cut down rosewood trees on the southern slopes of Masoala, trees that were many hundreds of years old. The poachers hacked pathways down to the beach, destroying hundreds of other trees in their path deemed worthless. The logs were then either transferred by small boats to "mother ships" anchored out of sight offshore or openly loaded into containers in Maroantsetra under false cargo manifestos, the so-called yellow-channel. Massive hauls of illegally shipped rosewood were made in Singapore, Mombasa, and Sri Lanka. One confiscation in Singapore alone netted thirty thousand rosewood logs plundered from the forests of Madagascar following a coup d'état in 2009.

In our weeklong stay in Masoala, we saw one small group of very skittish red-ruffed lemurs deep in the forest. Hardwood poachers had eaten the rest.

In the coup d'état of 2009, the young disk jockey Andry Rajoelina, who was the mayor of Antananarivo, ousted President Marc Ravalomanana. Scores of people were killed on the streets of Antananarivo. Ravalomanana took refuge in South Africa in the trendy beach resort of Umhlanga. Andry Rajoelina was declared head of the High Transitional Authority, a stand-in mob dominated by members of the Determined Malagasy Youth Party,

Rajoelina's party. A proper government was finally elected democratically only in 2013.

The coup d'état in Madagascar saw the immediate emergence of re-source-plundering Chinese syndicates in cahoots with influential members of the High Transitional Authority—vultures intent on plundering the set of the greatest biodiversity show in the world. The poachers targeted the forests in particular for hardwood: rosewood and mahogany for furniture—very expensive furniture—in China. There was nobody to stop them. Masoala National Forest was hit particularly hard, as were the red-ruffed lemurs, which were hunted by the Malagasy workers to supplement the daily rice rations dished out by the Chinese exploiters.

I left Madagascar because I had teaching commitments and left Keri and Danielle behind in Antananarivo. I was angry. I felt that I had been robbed of a special moment in my career, the thrill of establishing my first research camp on foreign soil. I was also incensed with the Chinese, who were so quick to exploit a political weakness to plunder the resources of Madagascar.

Keri and Danielle established the camp without my help after the equip-ment was eventually released. We did not pay one ariary in bribes in the end. It cost me in hotel bills for our unintended stay in Tana, but at least we did not submit. I refused to use my precious research funds to support corruption.

And then came the next surprise.

"*Merci, madam*," the Malagasy head of research at Ankarafantsika in-formed Danielle, "you cannot enter the forest at night."

The Malagasy "government" had recently issued a decree forbidding *va-saha* from entering any national park at night. They did this to spare them-selves potential ugly international incidents in which foreign tourists might meet up with the poachers in the forests at night. Keri and Danielle were armed with a permit to work on two species of nocturnal animal in the for-est in Ankarafantsika, but they were not allowed to enter the forest at night.

Keri and Danielle had to wait another two months to get special permis-sion to enter the forest. Two months is a long time when you live in a tent in the hot, sticky tropics unable to do anything constructive.

By the time they could start catching tenrecs, it was March, three months since their arrival in Madagascar. Danielle focused on *Setifer* and Keri on *Tenrec*. They worked with Ndrema. He accompanied them every night on their trapping trips into the forest. Keri and Danielle soon discovered that it was very difficult to trap tenrecs with baited walk-in traps. There were no walk-ins. They resorted instead to hunting by hand under the expert tutelage of Ndrema. Keri explained their technique in the *Live Science* interview: "When we heard the characteristic rustle of a tenrec in the surrounding forest, my guide or guides would circle the animal and flush it out onto the path where I could either sneak up to it or grab it quickly . . . the renowned poor eyesight of tenrecs definitely worked in my favour."

Before any implants were performed, the animals were fitted with collar-mounted radio transmitters to test our tracking techniques. The collars all fell off within 24 hours, discarded on the forest floor. Tenrecs do not have necks as we know them, that is, sections of their upper torsos that are narrower than their heads and shoulders. They have V-shaped torsos from their rib cages forward.

Then one of the most important pieces of our equipment stopped working. An electrical component in our field oxygen analyzer could not cope with the heat and the humidity, we think. The humidity at Ankarafantsika was awesome. It condensed as water droplets that ran down the walls of the huts. Fungus grew on my camera lenses.

We were getting nowhere. We could not track the animals, and we could not make the measurements of metabolic rates that were essential to understanding endothermy. I was keeping track of events at home with growing trepidation. How were we going to solve these problems?

In my laboratory at the university I spent many hours fiddling with radio transmitters, trying to figure out how I could implant them *into* instead of *onto* the tenrecs. The solution was really quite simple in the end. I wrapped the whip aerial around a 10-mm bolt and secured the coils in place with heat-shrink tape. Then I unscrewed the bolt from the aerial and so converted the whip aerial into a coil aerial. I figured that I could package the converted radio transmitter with two temperature loggers, encase the whole thing in surgical wax, and implant it as one unit.

I organized a replacement oxygen analyzer to be sent from the manufacturers in Las Vegas. They bent over backward to help us out. As soon as it arrived, I made plans to go on a rescue mission to Madagascar, in the middle of my teaching block.

A hot, humid breeze was blowing off the sea as I walked across the tarmac toward the airport arrivals terminal in Majunga. Majunga is a coastal port on the broad estuary of the Betsiboka River, about 120 km from Ankarafantsika. Keri and Danielle were there to meet me, beaming from ear to ear, but looking battered.

By the end of the day, I, too was battered after my first *taxi-brousse* ride in Madagascar.

Danielle had pre-negotiated three spaces in the 2 p.m. taxi-brousse leaving for Antananarivo via Ankarafantsika. As the *vasaha be*—the big foreigner—I was mercifully allocated the passenger seat, a special honor from what I understood. I still do not know for whom the passenger seat is routinely reserved. Keri and Danielle were wedged among the masses in the back seats.

There are many scary things about taxis-brousses, such as some of the passengers—one licked Keri on the arm—the dilapidated state of the vehicles, or the level of inebriation of the drivers. But the scariest of all was the stuff stacked on the roof—lots of stuff, too much stuff—along with my luggage. Barrels of paraffin and cooking oil, buckets, chickens, stitched-up woven baskets, boxes, bulging hessian sacks with moving things inside, bicycles, and motorbikes were all stacked three or four meters high. On every corner the taxi-brousse leaned and groaned, bent and tortured from too many journeys. I clung to my dashboard handrail. As night approached, we stopped in a small village to buy boiled duck eggs from a roadside hawker. Soon everyone was eating smelly boiled duck eggs, except me. I just wanted to get there.

The heat and humidity were shocking, even at around 8 p.m., when we were finally regurgitated from the taxi-brousse at the entrance to Ankarafantsika National Park. I was allocated a hut adjacent to the research camp. Its interior was painted with pictures of snakes, geckos, and lizards, all weeping

beads of condensation. The mattress and bedsprings were trashed from decades of gone-to-seed fornicating Frenchmen. I could not sleep. I lay on my
back, which I never do when I sleep. It was the best way to cool my chest.
The sweat pooled in the valley of my chest, a literal pond of sweat.

In the early hours of the morning, a creature walked across my chest toward the pond. I slapped it in the dark, very hard—splash, squirt! I reached
for my headlamp to see what it was. It was a spider, a huge spider, ham-
burger sized. I hate spiders.

I was very grumpy the next morning when I got up covered in spider
juice.

There was a quaint open-air restaurant at Ankarafantsika that fed tourists
three times a day. The tourists have an à la carte menu. The restaurant also
provides researchers with a fixed, take-it-or-leave-it dish every day at a third
of the price. The dishes are not varied—chicken and rice, beef and rice, or
tilapia and rice. You can, of course, order rice and rice, which many of the
locals do—huge, steaming bowls of the stuff. A tilapia is a small freshwater
fish that lives in the rice paddies and waterways. Chicken is a long-legged,
tough bird highly adept at dodging taxis-brousses. Beef is zebu, the scrawny
Madagascan ox.

But the restaurant also makes French fries, the best French fries on the
planet. Not those skinny, spiky things that you get at McDonalds; big, fat,
crisp wedges of pure potato freshly fried in the deepest, hottest oil. When
made under your own hungry gaze on an open fire, they are the best germ-
free nosh in Madagascar. At that time in my life I did not care a hoot about
overeating carbohydrates. I persuaded Lana, the waitress, to convince the
cook to fry me up *pommes frites* at any time of the day. Lana was my savior at
Ankarafantsika. She fed me and brought me beer—Three Horses Beer—ice
cold, at odd hours.

I spent the next few days teaching Keri and Danielle how to assemble
my new radio transmitter arrangement. I had to teach them how to solder
very small things together, delicately. They were great. Then we created the
package, the modified transmitter, waxed together with two temperature
data loggers. We were ready to try the first implants. All we needed now
were tenrecs.

On most nights I accompanied the hunting gang into the forest—Danielle, Keri, Ndrema, and a few guides. We each carried a stick with a forked end, big enough to pin a tenrec to the ground and get a glove on it before it could shred our hand. *Tenrec* is a vicious little creature that can inflict a very nasty bite. *Setifer* is just cute. It rolls into a ball, like a hedgehog.

After two days I developed a sweat rash on my back from wearing a wet shirt throughout the day. The itchiness was worse than real pain. Keri and Danielle tried everything to help me, but they could not find a cure. In desperation I asked them to apply Friar's Balsam that I found lying around in the camp. The pain was shocking. I goose-stepped down the N4 that runs through Ankarafantsika screaming silently.

And then they caught a tenrec. I had agreed to do the first surgical implant. After anesthetizing the animal, I scrubbed up and sat down to do the implant. Keri had to rub my back while I was working, but I was jerking, involuntarily. Suturing the animal took an eternity. I had to have full hand control for each stitch, each breath-holding, teeth-gritting stitch. Between stitches Keri rubbed by back and Danielle wiped the sweat from my brow. I was a very pathetic human indeed, but our first tenrec had been implanted. We went to Lana to drink Three Horses Beer.

I solved all of the problems within a week. I also finally threw off the skin rash—that was an enormous relief. The solution was talcum powder.

My return flight was in two weeks' time, and I had become redundant in the research camp. When students don't need your help anymore, they let you know very quickly—not directly, but subtly, through transmuted behavior. You get in their young space and breach their alliances, so they start to ignore you. Headphones appear. Living cheek by jowl in the small research camp required a how-to-live-with-students etiquette that I had to master fast. I needed to de-professor myself, and I was not too sure how to do it. There were other research students in the camp—Americans, British, and Japanese—all sharing the cramped spaces under the open-air thatched lean-tos. I had to learn how not to notice them in the mornings when they were grumpy. I had to learn how not to make comments about the weird stuff that they ate, the weird music that they listened to, and the weird things that they talked to each other about. I started to feel rejected, which is an awful thing for me.

I found a book lying around the camp, *The Song of the Dodo* by David Quammen,[2] and went to Lana to read, eat pommes frites, and drink Three Horses Beer. I found a nook where every so often the faintest breeze puffed. Sweat dripped from my forehead onto the pages of the book but I had a small towel for dabbing. My spot looked out over a steaming shallow swamp that harbored the odd crocodile and turtle. A small herd of zebu cattle waded across the swamp defecating on everything, scattering herons and annoying the resident pair of noisy jacanas. It was breathlessly hot.

And there I started to read a book I'd been meaning to read for years.

Quammen's writing captivated me from the first paragraph. I was terribly impressed that a journalist would have the guts — the temerity — to try to popularize a theory as potentially boring to the layperson as the Theory of Island Biogeography. Not only did he get the science across exceedingly well; he did so also with stunning insight, wit, and humor.

After reading for a while, I stopped to ponder my moment. One thing was for sure — despite the heat, it felt good. I was reading a book and was really enjoying every second. I *never* get to read books at home. I get to drink beer, sure, because you can multitask when drinking a beer, but I do not read books. I simply cannot find the time to do so. Meanwhile my emergency stand-in lecturer was coping with my baby, my BIOL 324 module in evolutionary physiology, that nobody else had ever taught other than me, for twenty years.

I wondered how my "line manager," Jeff, would react if he knew what I was doing. He's a nice guy, a long-term colleague, but a sad victim of the corporatization insanity that was ripping apart my university and many other tertiary education institutions worldwide. Corporatized universities are not meant to harbor within them senior academics who get paid a salary to drink beer, eat Lana's chips, and ponder the warmness of tenrecs. Modern academics need to earn their keep by solving people problems.

Umphgrrrr, umphgrrrr, umphgrrrr! A troop of brown lemurs made their way across the swampside forest meters away from my face going somewhere in the heat of the day. Brown lemurs are always grunting at one another and are always going somewhere throughout the day. They also, somehow, look ridiculously guilty as they do so.

I read on.

Island biogeography concepts started flooding back to me from my student days.

When animals colonize islands that have yet to be colonized by other animals, certain predictable changes take place in the new species that evolve from the colonizing ancestors. These changes provide some of the best insights into how evolution works. For example, island colonizers change size; some get bigger, becoming giants, and some get smaller, dwarfs of their ancestors. Good examples come from the Mediterranean islands, extinct dwarf elephants and gigantic rabbits. Small colonizers get bigger because there are no or few predators, so they can expand on their ancestral home range sizes and become bigger. Big colonizers get smaller because they do not need to defend themselves against predators but, perhaps more importantly, because islands have limited and often-unpredictable resources. A smaller body uses fewer resources.

Did this happen to the tenrecs? Not really. It certainly did not happen to them as radically as it did to the lemurs that colonized Madagascar before the tenrecs. The tenrecs may have been restricted in their capacity to change size by their insectivorous diet. With the exception of ant and termite specialists, no insectivorous mammal throughout 230 million years of evolution has ever become large, let's say, bigger than 10 kg (22 lb).

The Bornean seafarers who landed on Madagascar about two thousand years ago would have beheld before them a staggering variety of lemurs, from tiny, mouse-sized ones to giants weighing 160 kg (353 lb), as much as silverback gorillas.[3] Today the mouse-sized ones still exist, the mouse lemurs, but the giants do not; they were all eaten by members of the first nation. The Asian colonizers also ate the biggest bird that ever walked on earth, the elephant bird (Aepyornis maximus). They ate it to extinction, along with a pygmy hippo.

Targeting the big stuff was not unique to the founding Malagasy. All human colonizers of new lands hunted the biggest animals first, and why not? The return for the effort was good—lots of meat, skins, and bones—and the risks were low. These animals had not evolved side by side with humans and did not perceive them as threats or dangerous predators. This has happened everywhere on the planet except Africa, where humans evolved. The first colonizers of Madagascar would have been able to walk right up to *Archaeo-indris fontoynontii*, giant lemurs that looked like the equally extinct giant ground sloths of the Americas, and plunged spears into their hearts. The animals would not have batted an eyelid. Many island creatures to this day do not bat an eyelid at humans. Visit the Galapagos as Darwin did and you will discover this for yourself.

No continent or island was spared the plunder once *Homo sapiens* departed Africa. Australia was hit first, about 45,000 years ago, followed by Tasmania 41,000 years ago, and then Japan, Europe, and the Americas. Today we call these the megafaunal extinctions—the extinctions of the big animals. Some biologists argue that climate change and not humans caused the purge of the big mammals. Others argue that both were to blame. There is no doubt that the frequency and increased severity of the ice ages were difficult for herbivore populations to embrace at that time as habitats shrunk and expanded, so climate change may indeed have contributed to the decline in population sizes. But climate change cannot explain the slaughter of Madagascar's megafauna. Hunting elephant birds was simply more rewarding than hunting mouse lemurs, and probably more fun: it could be done during the day.

Island colonizers also lose their defenses. For example, the ancestor of the Mauritian dodo was a pigeon. The ancestral pigeons flew to Mauritius from the African mainland, probably unintentionally, in a storm. Once they arrived, they found no other pigeons. They could feast on the food that pigeons eat without competition. One newly evolved Mauritian pigeon specialized at feeding on the ground, not in trees, so the ability to fly was lost. There were no pigeon predators. Why would a pigeon need to fly if the food was on the ground and there were no predators? And it could get bigger, and fatter, to compete with other fatso ground-dwelling proto-dodos. The life of the dodo would have been ideal—that is, until the arrival of humans, hungry seafarers, tired of eating the rubbish onboard ships. For more, read *The Song of the Dodo.*

Changing body size and losing defenses were only some of the radical evolutionary changes that are known to have occurred in island colonizers.

The emphasis in Island Biogeography Theory is pervasive: change, change, change!

I put the book down and looked up into the trees within meters of my nook, and pennies dropped, lots of them. All the parts of the puzzle that I had been working on my whole career fell together, in seconds, right then and there, next to a steaming swamp. Thank you, David.

I grabbed the book and ran, leaving behind my beer and Lana's chips. I never leave behind a beer, but I did then. I ran to the research camp. The girls were startled; nobody runs in the heat in Madagascar.

"Get me a pen and some paper," I demanded, and they did.

In a few minutes I had mapped out a model in a sort of time-flow diagram. I finally knew what I was doing or should be doing in Madagascar. I had mapped out a framework into which I could put the bits and pieces of the endothermy puzzle. What came to me in my penny-dropping hullabaloo was the realization that island colonizers *do not have to change* if there is no evolutionary driving force for them to do so. It was as simple as that.

Céline Poux from Radboud University in the Netherlands and her colleagues estimated that a tenrec first colonized Madagascar from Africa perhaps as early as 45 million years ago.[4] This colonization event would have occurred when the earth was still very warm and tropical, in the mid-Eocene. However, as the Cenozoic unwound, some continental parts of the planet cooled dramatically, leading to the diversity of mammals that we see on earth today. But Madagascar did not change that much. Apart from the appearance of some desert areas in the southwest, in Tandroy and *Echinops* country, it stayed tropical. Also, it was busy moving northward, toward the tropical latitudes, as was Africa.

Why should the features that we were looking for, features of endothermy, be any different now than they were when the colonization event took place? Well, maybe they are not. Maybe what we see in tenrecs today are the same characteristics that existed in Paleocene and Eocene mammals, perhaps even in Late Cretaceous mammals. These characteristics may have been preserved on islands that were not subjected to global cooling. Tenrecs may be living physiological fossils, living counterparts of Schrëwdinger, molded by the Late Cretaceous climate but very little thereafter.

The capacity for evolutionary change in body temperature in mammals was probably tempered during the Cretaceous because, as I will argue in Chapter 17, the scrotum had yet to evolve and the body temperature needed to be kept at a temperature that was best for sperm maturation and storage. However, endothermy was given free reign to evolve in the Cenozoic following The Day of Reckoning (discussed in Chapter 10). Apart from explaining how ancestral endothermic characteristics in tenrecs may have been retained through resistance to evolutionary change, I could now perceive how mammals might respond to climate change. I could now finally better understand my earlier work on the variation in the heat of mammals and birds. I could explain why there was a 10°C (18°F) difference in body temperature between a naked mole rat and a pronghorn.

In those mammals and birds that could afford it—that is, in those in which the fitness benefits of being hot outweighed the costs—there would have been strong selection to increase body temperature. But I leave this discussion for Chapter 16, because we are far from being finished with tenrecs at Ankarafantsika. Understanding endothermy in tenrecs, the mammals that physiologically might best mirror Schrëwdinger's mystery state, was fundamental to understanding why birds and mammals are hot.

What happened to me beside the steamy swamp eating Lana's pommes frites was that I could finally bundle all of my work under one comforting umbrella question: why did endothermy evolve in birds and mammals? My gut feeling had taken me, and the last of my research funds, to Madagascar because I needed to know why tenrecs were so weird physiologically. But now I was convinced, fortuitously or otherwise, that I was in the right place studying the right animals, but at the wrong time, in the middle of a coup d'état. Luckily my excitement dragged Keri and Danielle along. It injected in them the energy and emotional strength that they so desperately needed to continue slogging it out at Ankarafantsika.

And so I went home, again.

FOURTEEN

Ancient Hibernation

I started to get emails, difficult things to create and send in Ankarafantsika. To get a signal, Keri and Danielle needed to walk to the highest part of the forest, with a laptop. Getting emails from Madagascar was like getting handwritten letters in the postbox, something rare and special. But the emails were not good. Money was running out, for Keri and for myself. But worse than that, the implanted tenrecs were being eaten, by poachers, feral dogs, natural mammalian predators, and, most frequently of all, by snakes (Figure 14.1).

There are two species of boa at Ankarafantsika, the tree boa and the ground boa. Both eat tenrecs. We were amazed at how easily they consumed the spiny *Setifer*. The quills seemed to offer no resistance at all. After a few days the snakes would puke out the indigestible stuff of the tenrec, the quills and the radio transmitter package. The girls would find the snake before the puke. But what could they do except track the radio transmitter in the tenrec, inside the snake? And *Tenrec*, despite having a much larger body size than *Setifer*, seemed to get eaten more often. Indeed, by the end of the study in Ankarafantsika, about 70 percent of all implanted *Tenrecs* had been eaten by boas.

Nevertheless, we finally managed to get meaningful data from three animals after an extraordinary commitment by Keri and Danielle. They helped each other in their respective studies. Of course they fought like hell with each other too, which they hid from me, but they got on with the job at

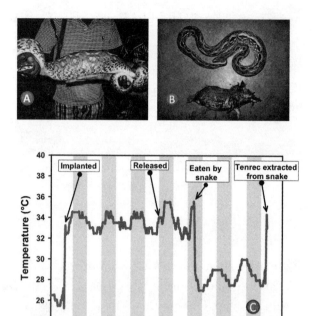

Figure 14.1. A Malagasy ground boa (*Acrantophis madaga-scariensis*), two days after it had swallowed an adult male tenrec (A). Boas caught and ate many of our study animals at Ankarafantsika, Madagascar. The tenrec was extracted from the live boa by carefully "milking" it out without harming the snake (B). The data logger implanted in the tenrec showed its body temperatures before it was eaten and when it was dead inside the snake (C). The regular humps in the data are the daytime elevations in body temperature, especially in the boa. (Data redrawn from Lovegrove, B. G., Lobban, K. D., and Levesque, D. L., "Mammal Survival at the Cretaceous-Paleogene Boundary: Metabolic Homeostasis in Prolonged Tropical Hibernation in Tenrecs," *Proceedings of the Royal Society B* 281 [2014]; photos by and used with the kind permission of Danielle Levesque)

hand. When one could not track her animals through malaria or other illnesses, the other stepped in to keep the work on track.

After the tenrecs had been implanted and released, three of them managed to avoid the snakes and burrowed into the safety of the sandy forest floor at Ankarafantsika. Keri and Danielle tracked the radio signals to the same place in the forest every day for nine months. The position of the signals never changed for nine months in three patches of sandy forest floor. Keri and Danielle were convinced that snakes had eaten the animals and that what they were tracking was a puked-up transmitter buried in the sand. But they could not risk digging the site in case this was not the case. What if the animals were intact? I pleaded with Keri and Danielle to be patient.

But then the radio signals started to get weird and irregular—battery failure! The risk of losing the data was simply not an option. The data would always tell some or other story, no matter what had happened to the tenrecs. It was time to dig down to the signal source. We needed those data, desperately.

Three digs, three tenrecs, in deep hibernation. Within an hour of being excavated, all three were ready to bite. There followed three surgical extractions and three nail-biting waits to see whether our data loggers in the packages had worked. The radio transmitters had worked for nine months, but what about the temperature loggers? After my data failures at Berenty I had developed a new distrust in the devices, the same devices.

We got data for nine months from each animal. We got perfect data—nothing missing, pure, beautiful, long-term body-temperature data: an evolutionary physiologist's opium. And the data were more stunning than we could ever have imagined.

All three tenrecs entered a state of hibernation as soon as they burrowed into the sand (Figure 14.2).[1] They effectively switched off. Their body temperatures plummeted to within about 0.2°C (0.36°F) of that of the soil temperature. Their body temperatures stayed like that for nine months. The temperature data loggers tracked the descent of soil temperatures into midwinter and the increase in spring, until we dug up the tenrecs. Not once, in nine months, did they reheat to a normal body temperature. These were the first data ever to show that a mammal can switch off completely for nine months without any apparent ill effects. No other hibernators are known to do this, although something similar has been found in dwarf lemurs, which I will discuss later. All hibernators reheat periodically to their normal body temperatures, say around 37°C (98.6°F), in what are called interbout arousals, and then sink back into hibernation within 24 hours.

The implications of the data were mind-boggling. Why was it that *Tenrec* and no other hibernator could get away without having to reheat

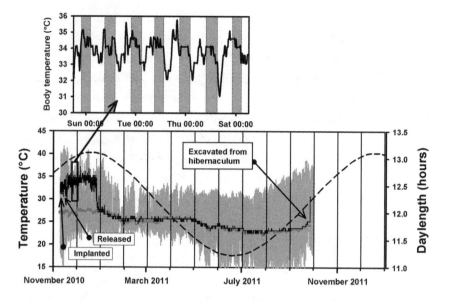

Figure 14.2. Body temperature data (solid line in the upper section) obtained from an adult male hibernating *Tenrec ecaudatus* at Ankarafantsika. The animal was implanted with a data logger and radio transmitter in early November and released a few weeks later. After a month it burrowed into the soil and stayed there for nine months. Its body temperature tracked that of the soil temperature (gray line in the lower section). The animal was excavated in late October when the batteries in the radio transmitter started to fail. Not once, in nine months, did the animal elevate its body temperature above that of the soil. The sinusoidal day-length curve (dashed line) emphasizes the small effect winter has on soil temperatures at a depth of about 250 cm (98.4 in). The light gray background shading is the air temperature at the site. (Data replotted from Lovegrove, B. G., Lobban, K. D., and Levesque, D. L., "Mammal Survival at the Cretaceous-Paleogene Boundary: Metabolic Homeostasis in Prolonged Tropical Hibernation in Tenrecs," *Proceedings of the Royal Society B* 281 [2014])

periodically? This is not a trivial question. Reheating from a hibernating body temperature is by far the costliest component of a hibernation bout. What is the point of employing energy-saving hibernation when it requires periodic arousals that consume huge chunks of the energetic benefits of hibernation? The answer undoubtedly will provide novel clues about the evolution of endothermy.

The answer to the question lies in understanding what happens when an endotherm cools down for an extended period of time.

Homeostasis is the intellectual playground of physiology. It is about keeping things constant, and much of mechanistic physiology concerns the study of how things are regulated within certain thresholds in the body. Irrespective of whether an animal is an ectotherm or an endotherm, big internal fluctuations in most things—temperatures, pressures, biochemical levels, and enzymes—spell disaster for homeostasis. The biochemistry of the warm body, for example, cannot tolerate fluctuations in temperature. It has much to do with how enzymes work because they work optimally at certain temperatures only. This does not mean that they should not work optimally at very cold temperatures. They can, and do, work just as well in Antarctic fish, which have constant body temperatures of about 2 degrees Celsius, as they do in pronghorns. What enzymes do not like, though, is a deviation in temperature from that at which they evolved to operate optimally. So hibernation body temperatures pose huge challenges for enzyme systems in endotherms.

Brian Barnes is a physiologist who works at the University of Alaska at Fairbanks. Brian recorded the lowest body temperature that has ever been measured in a live mammal, the Arctic ground squirrel, *Spermophilus parryii*.[2]

I met Brian for the first time in 1989 at Philipps University in Marburg, Germany—West Germany, at that time. Brian gave a research seminar in the Biology Department where I was doing a postdoctoral study with the doyen of non-shivering thermogenesis research, Gerhard Heldmaier. Brian showed us the most amazing data that I had seen up to that point in my career.

Barnes's laboratory had implanted squirrels with amplitude modulated (AM) radio transmitters, which were used in those days to measure body temperatures in small laboratory mammals. They were monsters to use. Because they were AM and not FM (frequency modulated), it was not possible to differentiate the signals from two implanted animals sitting, sleeping, or whatever, if they were side by side. Their worst characteristic was that they had a pathetic transmission range, perhaps a meter at the most. But Brian had managed to place antennae inside the hibernacula of hibernating squirrels, close enough to the animals to record signals throughout a complete winter hibernation spell.

Brian's signals reflected subzero body temperatures: minus 2°C (28.4°F) (Figure 14.3). I am not the slightest bit embarrassed to admit that the data give me a chill down my spine every time that I see them.

The seminar audience was stunned at first. How could a mammal hibernate in spells of three to four weeks at a time at subzero body temperatures without freezing solid? Mammals are, after all, 70 percent water, and water freezes at 0 degrees Celsius (32°F). Brian was talking to some of the world's experts on mammalian hibernation. Either his data were wrong or these animals had antifreeze in their blood!

Thomas Ruf was the first to raise his hand during the question time at the end of Brian's talk.

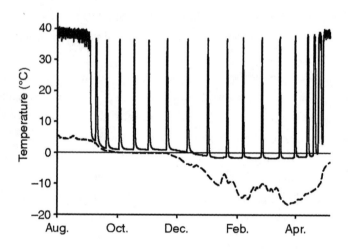

Figure 14.3. The most extraordinary hibernation data ever measured in a mammal: subzero body temperatures in the hibernating Arctic ground squirrel, *Spermophilus parryii*. The normal body temperature of about 38°C (100°F) was maintained during early August, after which the squirrel started to show bouts of hibernation lasting several weeks, interspersed with arousals lasting about 24 hours. Between January and April the body temperature during hibernation decreased to minus 2°C (28.4°F). (From Barnes, B., "Freeze Avoidance in a Mammal: Body Temperatures Below 0°C in an Arctic Hibernator," *Science* 244 [1989]: 1593–1595; printed with the permission of the American Association for the Advancement of Science, License no. 4031781110200)

"Brian," said Thomas, somewhat apologetically, "the operating range of these devices is from 0 to 50 degrees Celsius. Your measures fall outside the operating range. Any comment?"

Brian had none.

Brian repeated his experiments later with devices equipped to handle sub-zero temperatures and confirmed his original remarkable findings. He also confirmed that Arctic ground squirrels do indeed have antifreeze in their blood during winter: glycerol. Glycerol forms strong hydrogen bonds with water molecules by competing with the hydrogen atoms between water molecules, which disrupts the crystal lattice formation of ice. When a fat molecule is hydrolyzed, the products are one glycerol molecule and three fatty acid chains. By increasing the amount of a nonionic substance such as glycerol in the blood, the colligative properties of the blood are changed, importantly in this case, lowering the freezing point of water. What a biochemical convenience—using lightweight fat to store energy, using the fatty acids for fuel, and using the byproduct, glycerol, to lower the freezing point of the blood. Cool.

The Arctic squirrel data represent the evolutionary pinnacle of metabolic and endothermy flexibility that has ever evolved.

Not many mammal hibernators can achieve the low hibernating body temperatures of the Arctic ground squirrel. This ability is a highly derived, recent innovation in hibernators. On average, body temperatures get down to about 3.9°C (39°F) between arousals in hibernators. A reduction from a normal body temperature of, say, 38°C (100°F) to 3.9°C (39°F) represents a reduction of more than 32°C (61°F), nonetheless, a massive violation of homeostasis and a thermal shock to the biochemistry of the animal. What does it mean when a body temperature decreases so much?

One consequence of a lowered metabolic rate in endotherms is ischemic damage caused by reduced blood flow to tissues and hence an insufficient oxygen supply, especially in those mammals that are not equipped to deal with it. Humans, for example, are particularly pathetic when it comes to body temperatures that decrease by, say, 5°C (9°F) below the normal temperature. The brain gets all fuzzy, and the human heart is exquisitely sensitive to temperature. When the body temperature decreases to 30°C (86°F), the heart starts to fibrillate—it loses its normal pumping rhythm—

and becomes arrhythmic. Below 30°C (86°F), cardiac arrest occurs. *Homo sapiens* is not a natural hibernator.

Even although hibernators can avoid ischemic damage to their hearts, they do not seem to be able to avoid long-term biochemical ischemic damage: impairments of biochemical homeostasis. A reduction in oxygen reduces a cell's fuel supply, its concentrations of and access to ATP, the gasoline of the body. Molecules start to accumulate in higher-than-normal levels because they are no longer being converted to their normal end products. For the same reason, end products start to become depleted, reaching levels too low to sustain vital functions in the body. Biochemical accumulations and depletions can be lethal and must be avoided. They can be tolerated to a certain extent, but at some stage the imbalances need to be restored, and it seems that taking the body back to its normal body temperature for a fraction of a 24-hour day is sufficient to ensure this. The restoration of metabolic imbalances is the best supported hypothesis for interbout arousal during hibernation.[3]

So why do tenrecs not do this?

We do not know, and, as yet, there are no reasonable hypotheses either, except for one that I can offer, which has yet to be tested or evaluated by my peers. Allow me to explain it.

There may be a body temperature threshold only below which metabolic imbalances occur. The body temperatures of the tenrecs never once decreased below 22°C (72°F) throughout the long nine-month hibernation season. They survived the event, so clearly they did not incur pathological metabolic imbalances. If such a threshold exists, it need not necessarily be the same in all mammals. If tenrecs do indeed retain the ancestral physiological characteristics of endothermy, as I argue in this book, they may also have retained the capacity for a wider range of body temperature tolerance before metabolic imbalances occur. This makes sense if, as I argue, the earliest endotherms were adapted to being cold-blooded most of the time and ventured into warm-bloodedness at night only: nocturnal endotherms. Once mammals became more rigid defenders of a constant body temperature throughout the 24-hour day later in the Cenozoic, such a wide physiological tolerance would have been lost over time as enzyme systems adapted to operate optimally at a considerably reduced range of body temperatures.

NASA should be interested in these tenrec data because they have implications for sending humans into deep space. The current prediction is that astronauts will go to Mars within the next twenty years. To accomplish the nine-month journey—it would be a nutter's one-way ticket—the astronauts will need to travel in a state of induced hibernation to save on fuel, food, and water. Assuming that ischemic complications such as heart fibrillation and impaired brain function can be overcome, it is important to know what the lowest safe hibernation body temperature might be for astronauts. In Keri's tenrecs no ischemic damage seemed to have occurred when the animals hibernated at body temperatures at or slightly above 22°C (72°F). Would this threshold temperature be the same for humans? If it were, there would be no need to periodically arouse the astronauts during the entire nine-month trip. I foresee some interesting debates and developments over the next few decades in the context of this fascinating prospect. There has been intense interest in understanding the biochemistry of hibernation in *Tenrec* since the publication of our data. I helped Frank van Breukelen establish a laboratory colony of *Tenrec* at the University of Las Vegas, and the first data coming out of his research laboratory are extremely intriguing.

The capacity for a mammal to hibernate for long periods of time in the tropics is an idea to which scholars of hibernation are not accustomed. The study of hibernation has, in the past, been the preserve of North American and European biologists, who have studied it in animals that are adapted to extreme cold and high altitudes: bears, marmots, bats, and ground squirrels of all kinds. It is they who established and sustained the Holarctic Hibernation Paradigm. Cold habitats are relatively new habitats on earth, having appeared only within the past 30 million years. The hibernation characteristics that have evolved in the birds and mammals that colonized these new cold climes have shown highly controlled deviations from warmness, highly derived characteristics. They do not constitute the raw, ancestral state, such as we have identified in *Tenrec* or what we see in echidnas and some marsupials, endothermic states that evolved in a tropical world.

What we observed in *Tenrec* is likely to have been similar to what existed in Schrëwdinger when the meteorite slammed into Chicxulub. Schrëwdinger survived the impact, along with some ancestral marsupials, monotremes, and some mammalian lineages, such as the rodent-like multituberculates. I'd argue that Schrëwdinger survived because it could hibernate, for long periods of time, until the worst of the ecological disaster aboveground was over. It may even have been able to hibernate for a year or more because the soil temperatures would not have been low enough to induce long-term ischemic damage. It was a tropical world. At no time since the Permo-Triassic Mass Extinction event 252 million years ago was the earth ever as cool as it is now. What we see in *Tenrec* now may also be no different than the hibernation suggested for the cynodont *Thrinaxodon* 250 million years ago,[4] which I discussed in Chapter 4. Hibernation in mammals may be as old as the origin of endothermy itself.

Tenrecs are not the only mammalian hibernators on Madagascar; primates there hibernate as well. The ancestral lemur that colonized Madagascar from Africa might have looked something like the aye-aye (*Daubentonia madagascariensis*), because the family to which the aye-aye belongs, as its sole representative, Daubentoniidae, is at the root of the lemur family tree.

Céline Poux and her colleagues estimate that the colonization occurred as early as 60 million years ago, which is when the split occurred with the lorises, or around 50 million years ago, which is when the lemurs started to radiate on Madagascar.[5] The crossing of the Mozambique Channel occurred during the warming phase of the Late Paleocene and Early Eocene, when the earth was at its hottest in 300 million years.

There has been much debate about how mammals such as these managed to get to Madagascar: primates, tenrecs, rodents, carnivores, lipotyphlans (shrews), and hippos, which dwarfed after arrival. The two options are that the ancestor of each group got there either by rafting on floating vegetation or via an ancient land bridge. A new study argues that the cat-like carnivores (felids), got to Madagascar via an ice bridge with Africa following the dramatic global cooling during the Oligocene 30–27 million years ago.[6] The floating option, though, is the oldest and still seems to be the most favored idea, certainly for the colonization by lemurs and tenrecs. One of the arguments in favor of rafting is that only selected families within mammalian orders got there. For example, monkeys and apes did not get to Madagascar. Why did lemurs get there and not monkeys?

George Gaylord Simpson was the most influential paleontologist of the twentieth century. He contributed greatly to the modern synthesis of evolution and published several award-winning books with evolutionary themes. Perhaps his most famous treatise was on species dispersal. Here is the first published hypothesis by Simpson on how and why only certain groups of mammals got to Madagascar by rafting, or what he called "adventitious migration":

> Adventitious migration has indeed been used and sometimes abused simply to get inconvenient facts out of the way of a favored hypothesis, but there are instances in which adventitious migration is itself the most probable hypothesis and the most economical theory. In the cases of the faunas of Madagascar and the West Indies, for instance, I strongly favor this explanation, and I do so not at all in order to explain away data for a land bridge where I do not want to believe in one. . . . It is to be favored because it does explain, simply and completely, facts that the land-bridge theory does not explain . . . an adventitious route, which I call "a sweepstakes route" to emphasize this characteristic, is indeterministic. Its use depends purely on chance and is therefore unpredictable and, except in a broad way, can not be clearly correlated with other events in time and space, as filter-migration can.[7]

Civets got to Madagascar, but not cats or dogs. Lemurs got there, but not apes or monkeys. Mice got there, but not rats. Shrews got there, but not hedgehogs. Only certain families within orders got to Madagascar because they had drawn the sweepstakes ticket and the others had not. The rabbits and hares never got the ticket and were never in the sweepstakes.

The sweepstakes model of the "chance" colonization of Madagascar has been criticized due to the nature of the ocean currents between Madagascar and Africa, as well as the wind direction of the major cyclones. These both lead to the conclusion that a raft would not make it across the Mozambique Channel but would have washed up on a beach somewhere along the African coast.[8] But Jason Ali from the University of Hong Kong and Matthew Huber from Purdue University in Indiana have come to the defense of Simpson's famous hypothesis. They examined the ocean currents that existed at the latitude of Madagascar and Africa *during the Eocene* and found that these would indeed have enabled a raft of vegetation to reach Madagascar: "All signs point to the Simpson sweepstakes model being correct: ocean currents could have occasionally transported rafts of animals to Madagascar from Africa during the Eocene. Specifically, transport should have been from northeast Mozambique and Tanzania to the north coast of Madagascar. Given the slow tectonic drift of the island, this configuration probably continued at least through the Oligocene epoch. However, by the early Miocene, Madagascar breached the margin of the subtropical and equatorial gyres. Thereafter, currents were perennially directed westward towards Africa, making the ocean journey for mammals to Madagascar much more difficult, if not impossible."[9]

Ali and Huber also emphasize other aspects of the sweepstakes model that cannot be explained by alternative theories. Why is it that no large-sized mammals got to Madagascar? The hippo did indeed get there, but hippos can swim very well. Why did mammals such as elephants, rhinos, zebras, springbok, gemsbok, and eland not get there?

"Too big for a raft" is what Ali and Huber would argue, I'd guess. The raft would be, well, very wobbly if adorned with one or two elephants, no? It makes sense to me, although it should be pointed out that the non-afrotherian big mammals, the perissodactyls, such as the zebra, and artiodactyls, such as the eland, entered Africa only at a time when the currents had already started to become unfavorable for a channel crossing, as Ali and Huber suggest, "by the early Miocene."

The time that Ali and Huber estimate that it would have taken to make the crossing ranged from the slowest, about thirty days, to the fastest, about 5 days, the fastest on high-energy eddies. They would have had no food for five to thirty days; how did the rafters survive without food, being battered by waves and spray and strong winds?

In the case of primates, it was originally suggested that the colonizer was as small as a mouse or dwarf lemur and that, because these small primates are known to use daily torpor or to hibernate, they managed the crossing by hibernating in huddled groups in tree-trunk holes.[10] The problem with this idea was that it was assumed, for lack of more recent studies, that the ancestral lemur was a cheirogaleid, that is, a mouse lemur or a dwarf lemur. But we now know that this is not so; it is the aye-aye that anchors the lemur family tree.

I got tangled up in this debate when I was asked by Judith Masters—who, at the time, worked at the Natal Museum in Pietermaritzburg—to reconstruct the likely body size of the colonizing lemur using the most recent data on the family tree and body size of the lemurs that were driven to extinction by man a mere two thousand years ago.[11] I came up with a mass of about 1–2 kg (2.2–4.4 lb).[12] No lemur today that weighs this much or more is known to hibernate. Today's aye-aye does not, to our knowledge, hibernate, and by our reckoning, nor did its ancestor that made the first crossing of the Mozambique Channel. So why does hibernation ability suddenly appear in the mouse and dwarf lemurs? It was unquestionably a heritable characteristic, but it seems that the ancestral lemur was too big to have had the ability.

It is a bit of a mystery to me, because I would be extremely reluctant to argue that hibernation evolved independently in the cheirogaleids after the colonization event. The suite of genes involved in hibernation is simply too complex to have evolved twice independently in the placental mammals. Is it possible that the colonizing primate carried the genes for hibernation but that these were "switched off," perhaps through a process known as epigenesis, which seems to occur once mammals exceed a certain size?[13] When the smaller cheirogaleids evolved on Madagascar, were these ancient hibernation genes "switched on" again through dwarfing?

This is a question that geneticists can answer. However, it might help to look at the closest relatives of the Madagascan lemurs: the Lorisoidea, pottos, angwantibos, lorises, and galagos, or bush babies. It was from these primates that the lemur lineage branched off, so it would be highly informative to know whether these close relatives share the hibernation genes. None of these primates got to Madagascar.

The only lorisoid that approaches 1–2 kg (2.2–4.4 lb) is the potto, *Perodicticus potto*, which can attain a size of 1.6 kg (3.9 lb). Pottos inhabit the western African rainforests. The problem is that we are not sure about the hibernation capability of the potto. The earliest attempts to investigate thermoregulation in them merely stated that "at 10°C it may behave like a heterotherm."[14] This is a tantalizing statement but not very revealing. I believe that it is very important to establish whether the African prosimians do have the capacity for torpor and/or hibernation and that the potto is the perfect candidate in which to look for it because of its size.

I organized a project to do just this, but it was scuppered at the last minute by an outbreak of Marburg fever in the village that we intended to use as the base camp for the project. I had the money, the student, and the equipment. I'd even bought a clever anesthetic machine than generates its own medical-grade oxygen so that we would not have to lug heavy oxygen cylinders around. Marburg fever is Ebola's sister. You do not take chances with these lethal diseases. I pulled the plug on the whole endeavor and sent the money and the student to Borneo instead.

We have investigated whether hibernation or torpor exists in galagos, and there are two answers: no when the animals have access to normal food resources during winter, and yes when they do not.[15,16] Genes for torpor definitely reside within the Galagidae.

The other family within the Lorisoidea is the Lorisidae, the slow lorises. These primates occur in Southeast Asia and also carry genes for torpor and hibernation, but they are not bigger than 1 kilogram (2.2 lb). Until recently I would have argued that this family is in urgent need of study because there are no data available yet for free-ranging animals. So the story about how the cheirogaleids got their torpor and hibernation has yet to be fully told. Or so I thought.

Danielle sent me a new paper published in *Scientific Reports* on 3 December 2015 senior-authored by my friend Thomas Ruf and his colleagues.[17] It is titled "Hibernation in the Pygmy Slow Loris (*Nycticebus pygmaeus*): Multiday Torpor in Primates is Not Restricted to Madagascar." I was stunned. Thomas had found his data! I had visited Thomas where he works at the Institute of Veterinary Medicine in Vienna a few months earlier, in September 2015, and at that time he was seriously bemoaning the fact that a student had "lost" the data from the project on a train returning from Vietnam. We were having a beer at a beer garden a stone's throw away from the institute, and I can clearly recall the choice and unrepeatable German words that Thomas had reserved for this poor student. But the data were recovered, the student is presumably off the hook, and the results are exciting (Figure 14.4).

The study was conducted at the Endangered Primate Rescue Center, Cúc Phuong, Ninh Bình, Vietnam. Here is an extract from the abstract: "For true hibernation, the geographical restriction was absolute. No primate outside of Madagascar was previously known to hibernate. Since hibernation is commonly viewed as an ancient, plesiomorphic trait, theoretically this could mean that hibernation as an overwintering strategy was lost in all other primates in mainland Africa, Asia, and the Americas. Here, we show that pygmy slow lorises exposed to natural climatic conditions in northern Vietnam during winter indeed undergo torpor lasting up to 63 h, that is,

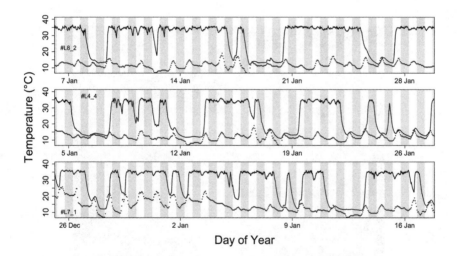

Figure 14.4. Nine-month continuous recordings of the core body temperatures of slow lorises, *Nycticebus coucang*, living in an outdoor enclosure at the Endangered Primate Rescue Center in northern Vietnam. Hibernation bouts occurred in December and January, when the body temperature decreased to about 12°C (54°F) for several days. (Data kindly provided by Thomas Ruf and replotted; reported in Ruf, T., Streicher, U., Stalder, G. L., Nadler, T., and Walzer, C., "Hibernation in the Pygmy Slow Loris (*Nycticebus pygmaeus*): Multiday Torpor in Primates Is Not Restricted to Madagascar," *Scientific Reports* 5, 17392 [2015])

hibernation. Thus, hibernation has been retained in at least one primate outside of Madagascar."

These data are important for two reasons. First, they confirm that there are non-Malagasy primates from which the cheirogaleid lemurs would have inherited the capacity for hibernation. Second, the idea that primate hibernation has been lost on mainland Africa, Asia, and the Americas supports my argument that plesiomorphic traits might be retained in regions, especially islands, that have remained tropical throughout the Cenozoic.

There is another alternative hypothesis as to how the lemurs got their torpor: aye-ayes may hibernate, or at least use daily torpor. These creatures are so poorly understood in the wild that this possibility cannot be dismissed until a serious attempt has been made to measure it. When I saw my first wild aye-aye in a forest near Maroantsetra, it was the massive size of the tail that made a first impression (Figure 14.5). Could it be that aye-ayes store

Figure 14.5. The rare and elusive aye-aye, *Daubentonia madagascariensis*. (Photo by Frank Vassen, http://flickr.com/photos/42244964@N03/3897947810; CC 4.0)

fat in their tails? We know that they build nests of leaves and twigs in the canopies of large trees and spend several days in their nests without emerging. It would not surprise me to document torpor in aye-ayes because they are such diet specialists. One of their favored food items is the nut of the *Canarium* tree. What happens when the trees are not fruiting, for example, during long, sustained, El Niño–induced droughts that plague Madagascar frequently?

Let's return now to the question of whether dwarf lemurs may also be avoiding interbout arousals? They live in exactly the same forests as the common tenrecs, so why should they not also display the unique long-term hibernation that lacks periodic arousals?

The answer to this question depends on where they hibernate. Dwarf lemurs (Figure 14.6) typically hibernate in tree holes, some of which are better insulated than others. Their body temperatures track those of the tree-hole temperatures very precisely during the hibernation season (Figure 14.7). Kathrin Dausmann from the University of Hamburg is an expert on dwarf lemur hibernation and has shown that in a well-insulated tree hole the tree-hole temperature and the dwarf lemur's body temperature fluctuate between about 10°C (50°F) and 25°C (77°F) daily.[18] Every four to five days, though, the animal will heat up to normal body temperature, about 37°C (98.6°F), for less than 24 hours, and then reenter hibernation. So this constitutes a typical periodic arousal superimposed on a daily cycle of body temperature driven by the tree-hole temperature. But if the tree hole is poorly insulated, the tree-hole temperature and the body temperature fluctuate daily between about 8°C (46.4°F) and 32°C (89.6°F), and the animal *never* arouses to its normal body temperature. Kathrin Dausmann has also reported that interbout arousals disappear if the daily body temperature reaches a rewarming threshold daily: "If T_b [body temperature] is raised above 30°C passively regularly, these individuals dismiss the regular active arousals and conform T_b passively over many months. . . . *C. medius* are the only hibernators that are able to abandon the necessity of arousals during hibernation and stay in a hypometabolic state for many months (even at ambient temperature of over 35°C [86°F])."

Figure 14.6. A fat-tailed dwarf lemur, *Cheirogaleus medius*, coming down a tree trunk head first. (Photo by and courtesy of Kathrin Dausmann, University of Hamburg)

Yet again, Dausmann's studies illustrate how flexible and adaptive the abandonment of endothermy can be when it is not needed, that is, when an animal is not breeding.

Of course we now know that tenrecs do this as well, but in the case of the tenrecs the lack of interbout arousals comes about because the body temperature is never cooled below 22°C (72°F) and so, by earlier reasoning, perhaps no ischemic damage has occurred. In the dwarf lemurs, in both the well-insulated and poorly insulated nests, the body temperatures do decrease below 22°C (72°F) daily, but only in the poorly insulated nests does the actual air temperature passively heat the animal above 30°C (86°F) on a daily basis to mend the ischemic damage, thus obviating the need for periodic arousals. These data add to the call for an expeditious study of ischemia onset during hypothermia in birds and mammals.

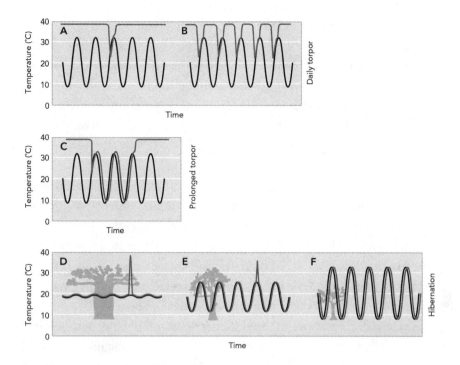

Figure 14.7. Stylized graphs based on field measurements of body temperature in the dwarf lemur, *Cheirogaleus medius.* The lighter line is the body temperature and the darker line the temperature inside the nest hole. Short bouts of daily torpor in which the body temperature does not decrease below about 23°C (73°F) are shown in A and B. The large daily range of nest temperatures suggests poor nest insulation. Prolonged torpor is shown in C, in which the lemurs allowed their body temperatures to track that of the poorly insulated nest for several days. Hibernation is shown in D, E, and F. The data in D show an animal hibernating in a well-insulated nest (low temperature fluctuations) in a large baobab tree. The animals' body temperatures track that of the nest, and both the nest and the body temperatures remain below 20°C (68°F). However, periodically the animals arouse (day 4) to normal body temperature, around 38°C (100°F). The same scenario occurs in E in a nest with lower insulation. In a poorly insulated nest (F), the lemurs' body temperatures track the nest temperature, but there are no periodic arousals because the animals are naturally heated to more than 30°C (86°F) on a daily basis. These lemurs store fat in their tails to fuel their long bouts of hibernation in tree holes. (Data obtained and kindly provided by Kathrin Dausmann and replotted; the original graphs were published in Dausmann, K., "Flexible Patterns in Energy Savings: Heterothermy in Primates," *Journal of Zoology* 292 [2014]: 101–111)

Both the dwarf lemur and the tenrec data show that tropical hibernation is not something unique or derived. I conclude that it is an ancient innovation that evolved in the tropics in concert with the evolution of endothermy when the first derived cynodonts and mammaliaforms were obtaining machinery for internal heat production for staying warm. If anything, interbout arousals may be the derived feature of hibernation, which evolved when the planet cooled within the past 30 million years.

Small primates on Madagascar use torpor and hibernation, but did other mammals that colonized Madagascar—the shrews, rodents, and carnivores—do likewise?

I'm not a gambler, but I'd be happy to put a few bob on the proposition that if we look for it, we'll find it. At the moment we simply do not have any data, so my bet is safe. The primates and the tenrecs have gobbled up all of the research money and attention on Madagascar.

There is one carnivore on Madagascar that would be my best bet. It is called a falanouc (*Eupleres goudotii*). The falanouc is a highly specialized carnivore; it feeds on earthworms that it digs out of the forest leaf litter or swamp edges. It weighs up to 4.5 kg (9.9 lb) and apparently can store up to 800 g (28.2 oz) of fat in its enormous bushy tail. Here is the kicker: dwarf lemurs also store fat in their tails before the hibernation season. Fat storage in tails is a telltale sign of hibernation.

I have spoken to many Malagasy on both the eastern and western sides of the island, and they all say the same thing: "Falanoucs go to sleep for two or three months in midwinter."

"Where do they go?" I have asked.

"*Tsy fantatray*," they have said; we don't know.

This completely anecdotal information is fascinating because it suggests that falanoucs might hibernate at the time of the year when earthworms are scarce.

Thomas Ruf and Fritz Geiser have assembled the largest database on the phylogenetic distribution of hibernation and daily torpor in mammals and birds to date.[19] In mammals it occurs in the Lipotyphla, Chiroptera, Carnivora, primates, and rodents. The Lipotyphla, the remnants of the old In-sectivora after the golden moles and elephant shrews were shifted to Afro-theria, has both hibernators (hedgehogs) and daily heterotherms (shrews). And indeed, so do the other four heterothermic orders, the bats, carnivores, primates, and rodents. So where heterothermy has been retained in a mam-malian order, it can exist in one of two forms, hibernation or daily torpor. The factors that determine the form that is employed by a species depends mostly on its body size and its latitudinal occurrence, that is, how far north or south it occurs.

I would argue that the hibernation that we saw in tenrecs and primates on Madagascar has not necessarily been influenced by these factors. Instead it has been molded by geographical isolation on a tropical island: isolation from a world that started to cool rapidly after the colonization events. I think that this argument can be applied to the new data for the slow loris in Vietnam as well.

Madagascar did not cool. It stayed tropical, and later semi-arid and arid zones appeared along its southwestern lowlands. From at least three million years ago it started to get whacked seriously hard by El Niño,[20] which re-ally made the seasons that are normally predictable highly unpredictable. So the capacity for daily torpor and hibernation was a perfect physiological asset to use to deal with these circumstances and can probably explain why those lemurs that are heterothermic are flourishing on Madagascar despite the best attempts by humans to drive them to extinction as they did their bigger counterparts two thousand years before.[11]

FIFTEEN

Twenty-Four Nipples

My research money was finished, but our work in Madagascar was not. Keri was particularly hard hit because, being a white student, she was initially, emphatically denied a bursary to undertake her MSc. At that time, the University of KwaZulu-Natal (UKZN) had decreed that no white students, whatever their financial circumstances, were to be granted bursaries. Keri comes from a broken home, and her family was most certainly not wealthy and could not afford to pay for her postgraduate education.

I got a phone call from Keri from Madagascar one day. She was in tears. "I cannot even afford to buy rice anymore, only bananas," she sobbed.

I had to do something, and I had to do it fast.

My savior was the chief technician at UKZN, Martin Hampton, who discreetly lent me money from a slush fund that he managed. It took me two years to pay it back, but I am forever thankful to Martin for his understanding, trust, and willingness to help. Fortunately, we could plan on an early exit from Madagascar while still completing the best of the research on the money that I had borrowed. I headed to Madagascar with wads of euros stuffed into my backpack. The first thing that I did was to get Keri off her banana diet. The guides got paid, and we paid off all other debts at Ankarafantsika.

Keri and Danielle noticed that the tenrec populations were dwindling, alarmingly. The local people were thriving on the intimidation of the park officials by the poaching syndicates and were resorting to their old hunting ways, using packs of dogs. It was getting very difficult to get data in Ankara-fantsika. I saw a taxi-brousse heading down the N4 toward Majunga with a cage of dozens of live *Tenrec ecaudatus* perched on top of the roof luggage. It was headed for Majunga's bush-meat market. The common tenrec, I suspect, will not remain common for very much longer.

Like Keri's tenrecs, *Setifer* (Figure 15.1) were also persecuted relentlessly by Malagasy ground boas.[1] We were puzzled that the quills seemed to be no deterrent whatsoever to the snakes. We reasoned that if the snakes made very fast strikes, while the tenrecs were active and not rolled into a ball, they would be able lay a few body coils around the tenrecs and suffocate them to death *before* they had a chance to roll into balls. I can't imagine that boas could swallow *Setifer* tenrecs once they were already rolled up into very spiky balls. Once the animals were dead, and provided that the snakes started swallowing at the head end, there should be no problem with the quills because they face backward, like the barbs on an arrowhead.

The fact that an ectotherm can prey on an endotherm at night does not quite fit my perceptions of what ectotherms should or should not be able to do at night, when it is cooler. Clearly, in the tropics, the nights are warm

Figure 15.1. The greater hedgehog tenrec, *Setifer setosus*. (The photo was taken by Danielle Levesque at Ankarafantsika, Madagascar)

enough to allow boas to employ bursts of speed and anaerobic metabolism to snatch tenrecs. They also have one very powerful hunting tool: they can see warm things at night using information collected by the infrared pits in their snouts. So ectotherms can and do hunt endotherms at night, and probably have done so throughout the enormously long period of coexistence of the sauropsid and synapsid lineages.

I was following Danielle's progress on the greater hedgehog tenrec *Setifer setosus* with immense interest. Although Keri's tenrec data had allowed us to speculate about how Schrëwdinger managed to survive the Cretaceous-Cenozoic boundary—namely, by abandoning endothermy for possibly as long as a year—Danielle's study was directed at evaluating one of the principal endothermy models, the Parental Care Model. We thought that we had a way of testing it.

Danielle's project was simple in theory but exhausting to carry out. We wanted to know whether these animals stepped up their use of endothermy when they bred. Colleen Farmer's Parental Care Model argued that endothermy should reach a peak during gestation and lactation.[2] To obtain the data that she needed, Danielle had to construct a two-part program. First, she needed to know the status of endothermy throughout the year during breeding and nonbreeding periods. Second, she needed to catch and bring *Setifer* tenrecs back to the research camp for a few days to measure their metabolic rates and body temperatures at different temperatures and at different stages during their breeding cycles. These tasks sound easy, but they were not.

When she started her study, Danielle, and everybody else for that matter, knew very little about the breeding and hibernation cycles of *Setifer*. All that she did know was what had been published up until that time, which was not very much. So the first year at Ankarafantsika was really just a long, hot, humid, learning experience for Danielle. However, being the doggedly determined Canadian that she is, she persisted for another two years. It was a very expensive, nonproductive year for the team, part of the risky game that we were playing to acquire unique data.

The data that Danielle finally did obtain from *Setifer* were very exciting and answered the questions that I had been asking since working with Fabien on *Echinops* at Berenty.

Setifer did not do the nine-month disappearing act that Keri's *Tenrec* did. This does not mean that they did not hibernate; they did, but for only about six months of the year. Remarkably, though, during the six months that they were not hibernating, some females managed to breed three times and still store a lot of fat for the upcoming hibernation period. This is a truly impressive reproductive output, and the pace at which it occurred was very rabbit-like: life in the fast life-history lane.

But rabbits are supraendotherms. They have some of the hottest bodies of all mammals, so we would expect that their gestation might be quick and that their babies might grow fast. *Setifer* tenrecs are not supraendotherms; they are basoendotherms. As it turns out, though, there is no relationship between a mammal's metabolic rate and how fast it grows.[3] There had to be another explanation.

The smoking-gun data came from the measurements of body temperature and metabolic rate when the females were in gestation and lactating.[4] This was the only time during the yearly cycle that they maintained a constant body temperature. This was the only time that they behaved like "normal" endotherms. The link between endothermy and breeding seemed to be paramount, which made sense to me, because breeding is the rationale of life, the mechanism of passing on genes to future generations. Natural selection was bound to favor any elevation in the thermal status of an animal if it boosted reproductive success. However, natural selection would also favor the retention of cold body temperatures if they saved energy during nonbreeding times of the year. This is what seems to occur in *Tenrec*, *Setifer*, and *Echinops*. They are essentially ectotherms when they are not breeding and become endothermic only when they breed. They are displaying exactly the characteristics that I speculated about in the mammaliaform, *Hadrocodium*, and possibly even in dogtooth as well.

The contrast between how *Tenrec* and *Setifer* breed is stark, yet the driving evolutionary forces that have led to the differences are exactly the same. There seems to be no doubt that the extreme predation pressure from snakes, in particular, has a lot to do with what can be seen in tenrec reproduction. *Setifer* is smaller than *Tenrec* and so can breed faster, because gestation times and weaning times are strongly determined by body size; they are shorter in smaller mammals. So *Setifer* can have several litters during the productive summer season. Danielle has shown that three litters can be raised during one summer. *Tenrec* cannot have multiple litters in a season; they are too big. They can have one litter only.

Both species need to counter the massive predation pressure from snakes by producing as many babies as possible during the summer: around 30 in the case of *Tenrec*, and 9 in the case of *Setifer*. These babies are "cannon fodder"—snakes will swallow most, and only a few will live long enough to breed. But if the predation pressures were the same on both species, *Setifer* would have less chance of persisting than *Tenrec* because it produces fewer babies during the year. So why does *Setifer* not have three litters of ten to match the *Tenrec* output?

Trade-offs, that's why.

Mammals pay a price for having many babies and many litters; they do not live long. Based upon its body size, we can predict that a one-kilogram female *Tenrec* should have a lifespan of about eight years if it was a "normal" mammal. Of course it is anything but normal, but we do not have any reliable data on its longevity in a snake-infested forest. It is highly improbable that a female tenrec could live for eight years producing 30 babies a year, a total of 240 babies in its lifetime. How would it manage to avoid the boas for eight years? If it had a 70 percent chance per year of being swallowed, as Danielle and Keri have observed, simple probabilities would show that it would be unlikely to live for more than two or three years.

There is only one evolutionary solution to beat the problem of high infant and adult death rates: breed early, and produce as many babies as possible as quickly as possible. This seems to have been the exact response that therapsids showed in the Early Triassic, although the extreme climate and not snakes drove their mortality. Tenrecs should reproduce as often as possible, before the snakes win the time race. This sounds logical, but there is one massive piece of the puzzle that was needed for the plan to work.

Tenrec do indeed have many babies; they have 24 nipples. *Tenrec* also breed as early as possible; they are very young at first reproduction. But what is puzzling is how a mammal with such an absurdly low metabolic rate and body temperature can have babies with such a fast growth rate. How can mammals with a low metabolic rate grow so fast?

The answer concerns energy-allocation priorities—investment "choices"—that maximize fitness and trade-offs. If a mammal does not need to allocate energy toward producing heat to stay warm—in other words, if it abandons endothermy—it can allocate the energy to growth instead if it is an infant, or to reproduction if it is an adult. In the adults, the fitness benefit of allocating energy to babies instead of heat production is that fecundity is maximized, that is, the total number of offspring per year can be maximized. The fitness cost, though, is the unlikelihood of sustaining this output in the years that follow. It is a case of breed fast and furious now, but pay a price later. But there may never be a "later" with so many snakes crawling around, so evolution will select for early litters of many babies.

Setifer may be "hedging its bets." Three litters per season is the maximum that it can produce, but the number of babies within a litter can be altered by natural selection. If having large litters reduces the likelihood of breeding in later years, it would be better to have a relatively small number of babies in each litter. *Setifer* is essentially hedging its bets against the likelihood of being eaten by a snake in later years. This interpretation implies, of course, that there may be differences in the mortality rates of *Setifer* and the common tenrec in the forests of Ankarafantsika. It implies that *Setifer* lives longer than *Tenrec*. The corollary to this argument is, of course, that *Tenrec* don't hedge their bets against being swallowed by snakes in later breeding seasons because over evolutionary time they have lost the bet too often. At present we do not have sufficient data to test these intriguing possibilities.

Of course why these tenrec scenarios do not happen in all mammals, by my argument, is because other mammals have not retained the ancestral condition of Schrëwdinger and her ancestors. Most other mammals have endured continental cooling and have not been isolated on a tropical island where maintaining a high constant body temperature throughout the year is not essential for avoiding the deleterious effects of hypothermia. Tenrecs suffer no physiological damage by abandoning a constant body temperature for nine months or more because the air or tree-hole temperature, and hence the hibernation body temperature, never decreases enough in winter to induce ischemic damage. But mammals that live on the continents need to maintain a high constant body temperature for the many reasons that we have discussed in other chapters: maintenance of the capacity for sustained locomotion, efficient lactation and gestation, defense of territory through-out the year, and so on. They cannot, for the most part, abandon energy allocation to keeping warm in preference to breeding. If they did that, they would freeze to death; they would go extinct, certainly at the higher lati-tudes. Also, they do not have to live with boas.

The results of the tenrec research condition cemented my suspicions about how, when, and why endothermy evolved in mammals. They pro-duce internal heat at the start of the night, become active, forage and feed, then flick the metabolic switch at dawn and slip down to whatever tempera-ture the day produces. When breeding, though, they avoid the dawn switch-flick sometimes because it retards gestation or lactation.

Our data that supported these conclusions were fascinating, but, as in the case of all enquiring research, they also generated as many questions as they may have answered. If this bouncing between ectothermy and endothermy was such a good evolutionary solution, why do birds and other mammals not do the same thing?

The answer to this question may have something to do with territory. When an animal disappears for nine months at a time, it is going to lose its territory, which means its access to food and mates. Those mammals that do not disappear maintain their territories throughout the winter, the normal hibernation season. If they do this, it simply means that there is enough food around for them to stay alive, but not necessarily to breed. Clearly this is not the case for hibernators, but it does mean that they can exploit extreme environments on earth: the very cold high latitudes, the derived global climate, and the unpredictable tropics, the ancient condition.

The data that we obtained from free-ranging *Setifer* have confirmed that what was originally observed in some tenrecs in captivity is actually happen-ing in the wild.

Peter Stephenson, who was then studying at the University of Aberdeen, was asking questions about the energetics of tenrec reproduction more than twenty years ago. His paper with Paul Racey on the large-eared tenrec *Geo-gale aurita* was highly revealing at the time that it was published, but I suspect that the data were too weird for their significance to have been integrated fully and fairly into energetic theories and hypotheses of the time.[5]

Geogale is tiny. It weighs less than 8 g (0.28 oz) when not pregnant. It looks like a small shrew with big ears. In several respects, Geogale is a remarkable mammal that can barely be called an endotherm. Like Setifer, it has an elevated metabolic rate only when it is breeding, that is, when pregnant or lactating. Otherwise, its body temperature simply tracks the environment's temperature, like that of a similar-sized gecko. Its body temperature has an influence on the rate of development of neonates because gestation can vary from fifty-four to sixty-nine days. It is also capable of postpartum oestrus, which means that it can mate immediately after giving birth and so can carry a litter while feeding the previous one: "The sustained maintenance of elevated RMR [resting metabolic rate] associated with the second consecutive pregnancy may not increase overall reproductive output for the breeding season, but overlapping the development of consecutive litters does result in reduced net resting metabolic cost. It also reduces the total time spent in gestation and lactation, which must reduce the probability of predation at this vulnerable time and increase the chances of survival for mother and litter. Therefore, in G. aurita, the ability to raise two litters in succession may be the main reproductive advantage gained from sustaining elevated RMR."[6]

The gestation length of Geogale is extremely long compared with those of other mammals of the same size. For example, the mean gestation length of a similar-sized lipotyphlan shrew, Cryptotis parva, is 22 days, compared with the average value of 61.5 days that Stephenson reported for Geogale. The long gestation time is a testament to the low metabolic rates of Geogale compared to the very high metabolic rates of shrews. This is an extreme, individual example of how metabolic rate can constrain growth rate during pregnancy through Arrhenius effects, but it certainly does not constitute support for a general relationship between metabolic rate and growth rate in mammals.[3]

When Peter Stephenson was doing his research on tenrecs in Madagascar in the early 1990s, hypotheses on the evolution of endothermy had barely progressed beyond the ideas of the 1970s. I have no doubt that, given hypotheses such as the Parental Care Model and the Assimilation Efficiency Model that we test now, Stephenson would have had a field day testing these models with his tenrec data. I evaluate these models in the last chapter, but I emphasize here that all of the laboratory data that Stephenson obtained in tenrecs support the Parental Care Model, in particular, stunningly well.

Several workers on tenrecs, including Stephenson, have reported that small tenrecs such as *Geogale, Echinops,* and *Setifer* enter torpor during gestation. In terms of the arguments that I have developed in this book, this general induction might require a reinterpretation.

Our default understanding of endothermy is that it is the ability of an endotherm to actively regulate an elevated, species-specific, constant body temperature. This we would consider the "normal" state for any endotherm, that is, the normothermic state, the state that you and I enjoy. Departure from normothermia occurs when an animal enters torpor following metabolic downregulation. So torpor is generally considered an adaptive departure from normothermia, often called adaptive hypothermia. But why can we not swap these metabolic states around? What if the normal state is ectothermy—hypothermia—in which the body temperature of the animal virtually tracks that of the environment, and the adaptive state is that in which the animal arouses to normothermia, a warmer, regulated state? This reversal of the default interpretation would be the plesiomorphic interpretation and would imply that tenrecs periodically enter endothermy, not torpor, by elevating their metabolic rate above that dictated by the Arrhenius effects on metabolism during certain phases in their reproductive cycles during, for example, lactation.

When they are torpid, variations in the daily air temperature, and hence in body temperature, can influence the rate at which a neonate develops during pregnancy, especially if there is no periodic entry into endothermy during the rest phase. Similar observations of lengthened gestation have been made in bats and some marsupials that enter torpor during pregnancy.

We have verified this pattern by observing carefully nonbreeding *Echinops telfairi* in my laboratory. We surgically implanted them with what is called a passive integrated transponder (PIT), generally called a PIT tag.

It is about the size of a grain of rice, and it returns a unique identification number and a body temperature when it is energized from the outside. PIT tags are dormant until activated, so they do not require any internal source of power throughout their lifespan. The tag is activated with a low-frequency radio signal emitted by a scanning device that generates a close-range electromagnetic field. The tag then sends a unique alphanumeric code and a body temperature back to the reader. PIT tags have become extremely useful research devices because they provide animals with a unique identity, a fingerprint, and also extra information, such as body temperature. They remain inside the animals for their whole lives, forever waiting to be activated to supply data without the animal's ever knowing that it is happening.

We placed *Echinops* inside a temperature cabinet every morning and observed their movements and behavior throughout the day using video cameras. For the most part they curled up into balls and, who knows, went to sleep. Their body temperatures in this state remained less than 1°C (1.8°F) above that of the cabinet. In more than 100 such individual procedures, not once did we see an animal attempt to maintain a body temperature remotely within range of those of other mammals during the rest phase. These tenrecs had flicked the metabolic switch. The only time that the metabolic rate and body temperature was elevated more than 1°C (1.8°F) above that of the chamber was when the animals were actively investigating their surroundings. Even then, the elevated temperatures caused by their activity did not approach anything close to what would be expected for a mammal of a similar size at rest.

These data were exactly what Fabien and I had measured in wild *Echinops* during winter in Berenty and confirmed that *Echinops* seems incapable of endothermy during the rest phase when it is not breeding. This is also the exact condition that I am proposing that the archaic mammals, the mammaliaforms, such as *Hadrocodium*, and perhaps the more derived, nocturnal cynodonts, such as dogtooth, would have displayed during the first stages of the evolution of endothermy following body-size miniaturization.

Stephenson and Racey could not have fully interpreted their data without knowing that snakes seem to be primarily responsible for the reproductive characteristics of their large-eared shrew tenrecs. Yet without knowing the real free-ranging situation, they were still able to invoke predation pressure as the likely selective force in the breeding pattern of *Geogale*. A high level of snake predation pressure seems to be the ubiquitous partial explanation for the unique physiology and metabolism of *Geogale, Echinops, Setifer,* and *Tenrec*.

SIXTEEN

The Pronghorn Pinnacle

To understand why some mammals are hot and others are cooler we also need to understand why different body sizes evolved during the Cenozoic—the dinosaur-free period of endothermy development. Schrëwdinger weighed somewhere between 60 and 240 g (2.1–8.5 oz), which is somewhere between the weights of a fat mouse and a skinny rat. Today's placental mammals range in size from the tiniest of all, the bumblebee bat, *Craseonycteris thonglongyai* (1.5–2.0 g) (0.05–0.07 oz), to the largest of all, the blue whale, *Balaenoptera musculus* (190 metric tons) (418,878 lb). The body sizes in placental mammals went ballistic once the dinosaurs went extinct, and it is astounding how quickly some mammals started to get big thereafter. During their approximately 140 million years of Mesozoic nocturnalism, early mammals remained small, or smallish, certainly compared to the largest modern mammals. But within a few million years of the Cretaceous-Cenozoic boundary, some mammals exploded from a few hundred grams to more than a million grams. Again, like the many other innovations that appeared in the Early Cenozoic, the dramatic increase in body size showed a huge pulse in the rate of evolution driven by remarkable circumstances at a remarkable time during the evolution of the mammals.

The relationship between body size and the pulses in endothermy during the Cenozoic is not obvious or well understood. It is true that the hottest mammals, such as cheetahs, hares, and pronghorns, are largish mammals, certainly compared with shrews, mice, and tenrecs. But the largest land mammal, the African elephant, has a relatively low body temperature (36°C) (96.8°F), and the smallest mammals do not necessarily have the lowest body temperatures. There is no straight-line, small-to-big, cool-to-hot relationship between body size and body temperature when all mammals are lumped together in the same analysis. The relationship is, however, strongly influenced by certain factors, such as where a mammal sits on the family tree (phylogeny) and how fast in runs.

John Alroy pieced together the changes that occurred in the body sizes of North American fossil mammals during the Cenozoic.[1] He estimated body sizes from mathematical relationships between tooth size and body size calculated from data obtained from living species. Using this comparative method, the body size of a fossil mammal can be estimated from a single fossil tooth. When Alroy plotted the data against time on the x-axis, going from about 80 million years ago on the left to the present on the right, and body size on the y-axis, a very curious pattern could be discerned. The starting body sizes were, of course, small, about the size of Schrëwdinger. The majority of mammals remained Schrëwdinger sized, or at least less than about a kilogram (2.2 lb), throughout the Cenozoic. But, as I have said, some mammals showed an immediate response to the demise of the dinosaurs at the Cretaceous-Cenozoic boundary—they got big.

Up until the onset of the Miocene 23 million years ago, some species of Perissodactyla, the mammal order to which horses and rhinos belong, were the biggest mammals that ever walked the earth. They belonged to the Indricotheriinae, which rivaled in size some of the medium-sized sauropod dinosaurs—when they were still around—and weighed over 20 metric tons (44,092 lb), more than double the size of the biggest African elephant (Figure 16.1). Known as hornless rhinos, they were forest inhabitants, browsers, and their fate was sealed once global cooling started to accelerate during the Oligocene and Miocene. The forests retreated to give way to open woodlands and savannahs, and these mega-mammals could not adapt to the new vegetation, so they went extinct. The Early Cenozoic was a period of spectacular evolutionary experiments in mammalian body size that has never occurred again. These experiments seemed to explore the very limits of body size in land-based mammals. Some large dinosaurs dominated the

Figure 16.1. *Paraceratherium* was one of the largest land mammals of the Cenozoic. It was a perissodactyl that went extinct 23 million years ago during the Late Oligocene. (Reconstruction by and courtesy of Dmitry Bogdanov, http://dibgd.deviantart.com, CC BY SA 3.0)

landscape before the meteorite impact, and it seems that some mammals filled these mega-beast niches very quickly once the dinosaurs went extinct. But, in the end, they too went extinct once the great rainforests dwindled.

By the end of the experimental tropical phase of the Cenozoic, body sizes in mammals settled down and followed two clear trends. Climate change was happening and, as always, started to drive extinctions and initiate new evolutionary patterns. Rabbits, hares, ungulates, and carnivores got bigger, and all other mammals remained small, less than a kilogram (2.2 lb), at least until the last 10 million years or so of the Cenozoic. John Alroy's graph shows this parting of body sizes very nicely. There was a steady, more or less straight-line increase in the sizes of mammals that can run fast starting at the end of the Eocene, that is, as soon as the earth started its rapid cooling phase. The parting of ways created a species gap: missing are species between 1 and 10 kg (2.2–22 lb), apparently a very unpopular size to be as a mammal.

Pinpointing the evolutionary driving forces responsible for the body-size pathways is paramount to understanding the diversity of mammals that evolved during the Cenozoic and therefore also the evolution of endothermic diversity. Selection for larger body sizes comes with great energetic and fitness costs. For starters, more food is needed to sustain a larger body size, which means that individuals need to look for food over greater areas. Herein lie dangers. Larger territories or home ranges expose animals to greater risks of predation. Large animals cannot easily hide from predators—they are persistently exposed and vulnerable. So, as always, we need to pin down the fitness benefits of being large.

My simple answer as to why animals got bigger during the Cenozoic is that they started to exploit the new dominant open-landscape plants: grasses.[2] Recall what we discussed in Chapter 2 about the problems involved in digesting cellulose. There I discussed these problems in the context of the stem amphibian ancestors of the "pelycosaurs," such as *Diadectes*, which were the first tetrapods to include plants in their insect and fish diet. There is a rule of thumb involved in the digestion of cellulose: endothermic herbivores cannot be too small. Of course many small animals such as insects also eat plants, and therefore cellulose, but they are not endotherms. Endotherms need to fuel the massively high costs of maintaining a constant high body temperature.

The more complex answer as to why mammals got bigger draws on some critical concepts in digestive physiology made explicit in a classic publication—to me, anyway—by Montague Demment and Peter Van Soest in 1985.[3] First, microbes need *time* to ferment cellulose. Second, the *rate* at which food passes through an animal is directly proportional to its body size. In small animals, the food passes though the digestive tract at a much faster speed than it does in larger animals. So, in small animals, there is simply not enough time for efficient cellulose fermentation to take place. There are some ways of getting around this problem, such as coprophagy, in which small mammals, typically, such as rodents and lagomorphs (rabbits, hares, and pikas), eat their own feces as they emerge so that the food undergoes multiple rounds of fermentation. But, for the most part, endothermic cellulose digesters need to be big. In mammals there is an optimal body size for the digestion of cellulose; it is between 80 and 180 kg (176–397 lb); the upper size is that of a wildebeest. It is hardly surprising, therefore, that wildebeest, together with the slightly larger zebras, form the largest herds on the grassy plains of East Africa. Ungulates are basically highly mobile fermentation vats, perfectly sized for the efficient digestion of cellulose. Small herbivores such as lagomorphs manage because they are obliged to include higher-quality vegetation with lower concentrations of cellulose in their diets.

Whether cellulose fermentation required or created high body temperatures in mammals remains a contentious issue. I take the view that the high body temperatures associated with many ungulates, such as pronghorns, have nothing to do with cellulose fermentation but everything to do with muscle power, running speed, and endurance. I make this argument from the standpoint that not all large-bodied cellulose fermenters have high body temperatures. For example, wombats and kangaroos are grazers, and they do not have high body temperatures. The average body temperature of a red kangaroo (*Macropus rufus*) is 36.3°C (97.3°F),[4] and that of a hairy-nosed wombat (*Lasiorhinus latifrons*) about 33.6°C (92.5°F).[5] The wombat is clearly a basoendotherm because its body temperature is less than 35°C (95°F), whereas the kangaroo is a mesoendotherm. These endotherm classifications show that a large basoendotherm *can* be an herbivore. The difference in body temperatures between these marsupial grazers and placental grazers is essentially involved in how they move. Kangaroos hop, and wombats waddle, whereas gazelles and pronghorns run—very fast.

Hopping, or saltatorial locomotion, does not require hot muscles. It is a form of locomotion that evolved in very energy-poor environments, and it minimizes the cost of covering vast distances to find scattered grasses.[6] The large hopping marsupials of Australia evolved in an environment in which there were no fast pursuit predators. There were predators, indeed, which were large ectotherms and some marsupials, but these were not fast. Unlike in the rest of the world, there were no placental carnivores in Australia. The marsupials evolved in isolation in Australia for perhaps as long as 80 million years.

The largest marsupial predator was the marsupial lion, *Thylacoleo carnifex*, the biggest mammalian carnivore ever to hunt in Australia. It was an ambush predator and could not chase after large herbivorous marsupials such as kangaroos. It weighed about 140 kg (397 lb), about a third of the size of a modern African lion, but it had a bite strength stronger than that of any modern mammal. It had enormously powerful jaws, long pointy canines for stabbing and holding prey, and extraordinary blade-like molars. It killed its victims very quickly, much more quickly than did African lions, which employed a tedious throat-suffocation process. It did not show any adaptations of the limbs that would suggest that it chased its prey. It hunted by pouncing on its victims from the branches of trees or by ambushing them at watering points.

After separating from Antarctica, a heavily forested, coldish Australia drifted north. As it did so, it got warmer and drier. The interior of Australia suffered the most; it became arid.

Most desert environments share one peculiarity in their rainfall: it is extremely unreliable and variable in time and space. The Sonoran Desert in the southwestern United States is one of the exceptions. Combined with isolation, rainfall unpredictability had a profound influence on evolutionary processes during the cooling phase of the Late Cenozoic in Australia. The large marsupial herbivores that evolved on the open, exposed grassy plains of Australia did not require speed to avoid predators, as placental herbivores on other continents did. This does not mean that kangaroos could not hop fast. They could, as the first Europeans discovered when they tried to run them down with their horses and foxhounds. Kangaroos acquired a form of locomotion that allowed them to travel vast distances to find scarce food resources cheaply. There is no known form of terrestrial locomotion that meets these demands better than hopping. Hopping breaks the rules of locomotor energetics, which we know from nonhoppers like you and me.

Imagine that I could persuade you to get onto a treadmill wearing a little face mask connected to an analyzer that would allow me to measure how much oxygen you used at different running speeds. The data would be highly predictable and would show the same patterns, but with different starting points and upper limits, for different people. Your oxygen consumption, an excellent proxy for your metabolic rate, would increase in a straight-line relationship with your running speed. The faster you would run, the higher would be the level of your oxygen consumption. But a limit is reached, not in running speed but in oxygen consumption. If there were no limit, you would be able to run as fast as you wished, perhaps 1,000 km/h (621 mi/h). The limit is called the VO_2max. It is simply the maximum rate at which oxygen can be delivered to your tissues, and it sets your maximum aerobic running speed. People with a naturally high VO_2max can run fast for long periods of time. These are the great long-distance runners of our time, athletes such as my heroes, Alberto Salazar and Bruce Fordyce, who inherited the ability from their parents.

Kangaroos have a VO_2max that does not set a limit on running speed. Indeed they show a straight-line increase in oxygen consumption during pentapedal locomotion, that is, using their tails in addition to all four limbs to move, but only as far as the VO_2max. When they start to hop, though, the line flattens out, so kangaroos can hop faster and faster with virtually no additional increase in oxygen consumption above their VO_2max.

Their capacity is associated with energy that is stored in their tendons and calf muscles during each bound, which is then released and used to propel the next hop. The mechanism is not unlike that in a pogo stick, in which the energy of each landing is stored in a spring and used to propel the stick upward on release of the stored energy. Kangaroos can cover vast distances at low cost, a highly useful attribute to have in the energy-poor, arid interior of Australia.

I believe that kangaroos have a lowish body temperature compared with those of other similar-sized placental grass eaters because they do not require hot muscles to hop fast. However, wombats are perhaps the best examples of large-sized grazers that have low body temperatures and metabolic rates. The resting metabolic rate of wombats is 36 percent lower than the average for other marsupials and a whopping 58 percent lower than that of a similar-sized average placental mammal. When you consider that similar-sized grass-eating placental mammals are not average because they have metabolic rates higher than other placental mammals, it is safe to say that wombats have metabolic rates less than half those of their placental counterparts. Yet they feed exclusively on grass, just like any antelope, zebra, or wildebeest.

Wombats walk with a waddle. At my fastest non-sprinting long-distance running speed I'd be able to keep up with a sprinting wombat going flat out for perhaps 10 minutes. We'd be doing around 15 km/h (9.3 mi/h). The wombat weighs about 35 kg (77.2 lb). But I would not be able to outrun a 35-kg springbok, which can reach speeds of 100 km/h (62 mi/h). The difference between the marsupial grass eaters and the placental grass eaters involves locomotion: how they move, their gait, their legs, and their speed. Wombats have short, powerful legs that have evolved to dig burrows. Indeed, this ability to spend long periods of time in a burrow with a low metabolic rate is probably what saved the remaining three species of wombat from extinction. Those wombats and close relatives that went extinct were the biggest grass eaters ever to have mowed the grasslands of Australia. The biggest wombat, *Phascolonus gigas*, weighed around 200 kg (441 lb), but the closely related diprotodons weighed a staggering three tonnes (3,000 kg) (6,614 lb). These marsupials were too big to dig burrows and would have suffered the consequences of fire, droughts, and hunting. They were driven to extinction very soon after the arrival of humans in Australia between 40,000 and 50,000 years ago.[6]

Placental grass eaters are hot because they run fast, but why did they steadily get bigger during the Cenozoic? Similar increases in body size occurred in the kangaroo and wombat lineages; their ancestors were also Schrëwdinger-sized forest dwellers. Apart from the gigantic *Diprotodon*, the biggest extinct kangaroo, *Procoptodon*, weighed about 240 kg (529 lb), more than a blue wildebeest, one of the most abundant grass eaters in Africa.

The answer to this question is, I believe, critical to understanding the evolution of mammals during the Cenozoic. Studies in the past thirty years on the evolution of mammalian body sizes have not given adequate recognition to Demment and Van Soest's classic paper. These agriculturalists showed that, to efficiently ferment grass, a mammal needs to be bigger than 10 kg (22 lb), optimally between about 80 and 180 kg (176–397 lb). Smaller than this and the grass eater will battle to meet its daily requirements through energy gained from grass fermentation alone. It will be forced to supplement its grass diet with plant foods having lower concentrations of cellulose. Larger than this and it will struggle to find enough grass in one day to meet its daily energy demands. It, too, will need to supplement its diet with better-quality food or grass by switching to browsing or both grazing and browsing, which is how elephants and the largest antelope, the eland, feed. Elephants don't push over trees for fun. They do it to access the new, tender leaves of the trees, which have yet to lay down structural cellulose. They are seeking out high-quality food, and it does not cost them much to knock down a tree. This is a very generalized perspective of how the smaller-to-larger scale works, but it is not of significant interest to my story to dwell on the possible variants of the patterns.

If there is an optimum body size range for converting cellulose to baby zebras and wildebeest, natural selection will, and did, drive the body sizes of grass eaters to this optimum and beyond, resources permitting, following the retractions of the forests and the expansions of savannahs, open wood-lands, and grasslands. Indeed the mammals that actually exist on the grassy plains of East Africa beautifully match Demment and Van Soest's theoreti-cal predictions about the optimum body sizes for cellulose fermentation. The Demment and Van Soest model is completely fundamental to any study on the evolution of body size in mammals, yet it is seldom taken into account, particularly by Holarctic macroecologists.

My principal argument is that the frequency distributions of mammal as-semblages on all continents except Australia represent a concatenation of the three principal locomotor gaits: plantigrade (a flat-footed heel-striking gait), digitigrade (walking on the toes), and unguligrade (walking on the toe tips): "The unimodal, right-skewed distribution, most frequently identified in contemporary descriptions of placental mammal body size distributions, masks an underlying multidistribution structure: a long-term evolutionary process that has generated a concatenation of two or three frequency distri-butions specific to locomotory modes. . . . Common interassemblage pre-dictions of such proportions in contemporary distributions may be disguised by the relative severity of the Pleistocene megafaunal extinction."[7]

Let me explain what I mean by "concatenation of the three principal locomotor gaits."

The plantigrade mammals are small. They include the rodents, shrews, moles, hedgehogs, and some afrotherians and xenarthrans. They all have flat feet—that is, they are plantigrade. They are also the most abundant mammals on earth. They show by far the biggest peak in numbers of species in a frequency distribution.

Imagine a bell-shaped curve with numbers of species on the y-axis and body sizes on the x-axis, centered on about 100 g (3.5 oz) and anchored at 2 g (0.07 oz) on the lower end and with an upper limit of 1 kg (2.2 lb). The vast majority of plantigrade mammals have stayed smaller than 1 kg (2.2 lb) throughout the Cenozoic for 66 million years. Seldom have flat-footed mammals evolved to body sizes larger than 1 kg (2.2 lb). If they have, they are primates, most of which live in trees; they are aquatic, such as many large South American rodents (e.g., capybaras); or they possess some form or other of body armor, such as armadillos. Flat-footed mammals, as I will discuss shortly, can also get big when there are no predators around.

The next bell-shaped curve is composed of digitigrade mammals, the carnivores, and the rabbits and hares. This bell curve is centered on about 25 kg (55 lb), the optimal size for a carnivore to be able to exploit both larger and smaller prey. It has a lower number of species at its peak than does the plantigrade bell curve. We'll discuss the rabbits and the hares shortly; they are quite special.

The third bell curve is composed of unguligrade mammals, the Artiodactyla and the Perissodactyla. This bell curve is centered on the optimal body size for herbivores, which we have discussed: the Demment and Van Soest optimum, say, about 80–180 kg (176–397 lb). Compared with the plantigrade mammals, this group also has a lower number of species at its crest. So in terms of sheer numbers, it is plantigrade first, followed by digitigrade and unguligrade. In terms of body size, the largest are the unguligrade mammals, followed by the digitigrade and then the plantigrade mammals.

For the African mammals, when these three bell curves are added together they sum to a bimodal distribution, a frequency distribution with two peaks that actually masks the underlying three peaks of the plantigrade, digitigrade, and unguligrade mammals.[7] Africa has the most impressive peaks and by far the most mammal species of all the non-Australian assemblages within each bell curve.

The intriguing question, though, is why does this underlying three-peak distribution come about?

I have argued that it is generated by the physiological optimal size for fermenting cellulose and the resultant predator-prey arms race.[8]

The body sizes of herbivores evolved toward the optimum of 80–180 kg (176–397 lb) because of the constraints involved in the fermentation of cellulose.[2] However, when these herbivores started to get big, at the end of the warm phase of the Cenozoic, let's say about 45 million years ago, the carnivores were still small, weighing, let's say, less than about 3 kg (6.6 lb). If the carnivores were to catch herbivores they needed to keep pace with the herbivore size spurts; they needed to get bigger and faster. They did not need to get bigger because of energetic constraints. They needed to get bigger because it was necessary to capture bigger prey—a clan of mongooses cannot bring down a Cape buffalo.

Bigger and faster carnivores spelled disaster for flat-footed mammals. To have flat feet carries with it the extraordinary evolutionary benefit of having fingers and toes. Try to imagine what life would be like without fingers. Try to imagine what it would be like to have two antelope-like hooves instead. Picking your nose would be out of the question. Try to imagine what it would be like to have paws with enormous claws. Playing the piano would be arduous. Seriously, though, plantigrade mammals need their fingers and their toes, for holding, manipulating, grooming, digging, climbing, and scratching. These activities were paramount for being able to exploit their local environments maximally. Digitigrade and unguligrade mammals cannot perform most of these activities. So a real problem emerged at the end of the Eocene for plantigrade mammals: they could not get bigger, as the digitigrade and unguligrade mammals were doing, so they could not have large home ranges and therefore could not forage for food over large areas. Unless they could fly, as bats do, migration to avoid cold, unproductive winters was out of the question. They had to deal with their local environments 24/7, 365 days per year. In a way, they were forced to become local specialists, which is indeed what happened.

Let's reiterate why plantigrade mammals could not get bigger than about 1 kg (2.2 lb). To run fast, mammals need long legs, reduced numbers of toes, fused fibula and tibia in their hind limbs, grossly elongated foot and hand bones, and hot muscles, as discussed in Chapter 17. These requirements rule out highly functional fingers and toes. So let's compare the maximum running speed of an average 10-kg (22-lb) plantigrade mammal with an average 10-kg (22-lb) digitigrade carnivore. I can do this because I have formulated the equations.[9]

The flatfoot would have a maximum running speed of about 25 km/h (15.5 mi/h) and the carnivore 50 km/h (31 mi/h), exactly double that of the flatfoot. Any theoretical plantigrade mammal that did manage to get to 10 kg (22 lb) and went out foraging in an open landscape where predators existed would have been the proverbial sitting duck; it would have been easy prey to a carnivore of the same size, bigger, or even slightly smaller. The evolution of open landscapes and the faster carnivores put the brakes on body-size evolution in all mammal groups that maintained the flat foot. That is why there is a gap in Alroy's graph.

A nice test of the flatfoot constraint is to see what happens to plantigrade mammals when they live in a predator-free environment. Consider what happened when the rodents first colonized South America.

Céline Poux and her colleagues, who estimated the colonization time of the tenrecs of Madagascar, also estimated that the ancestral rodents arrived in South America from Africa sometime between 45.4 and 36.7 million years ago, probably by rafting on a mass of floating vegetation that could have been washed out to sea from a West African river during a flood.[10] There were no large, fast mammalian predators in South America at that time. There were some small marsupial carnivores, but they, too, were flatfoots, so they posed no speed challenge. Dramatic speciation proceeded from this ancestral rodent into the diversity of caviomorph rodents that exist in South America today. By my reasoning, the South American rodents would have evolved to body sizes bigger than those that we see in rats and mice on other continents.

And that is what happened. By rodent standards, or even by general plantigrade standards, many of the South American caviomorph rodents are big mammals and far exceed the upper limits of the average bell curve of body size and the species numbers of all plantigrade mammals. They broke out of the flatfoot constraint. The ancestral rodents in South America arrived at a time when the earth started to cool rapidly, when forests retreated to make way for open landscapes. The caviomorph rodents could, and did, radiate into many niches yet to be occupied by the local fauna, mostly marsupials and the endemic, now extinct, South American ungulates, the Notoungulata. They radiated into aquatic, forest, woodland, savannah, grassland, and desert habitats.

The largest living South American rodent is the capybara (*Hydrochoerus hydrochaeris*), which weighs about 50 kg (110 lb). But the capybara is a midget compared to some South American rodents that are now extinct. There were several species that weighed about 580 kg (1,279 lb), the size of a large horse. The biggest rodent ever to have lived in South America was *Josephoartigasia monesi*, which weighed a whopping 1 metric ton, 1,000 kg (2,205 lb). The fossil was discovered in Uruguay and is about two to four million years old.[11] This age coincides with the time when South America was invaded by placental digitigrade carnivores from North America during the Great American Interchange. North America and South America were now linked by the Isthmus of Panama, which rose up from the seafloor. The North American carnivores included cougars, saber-toothed cats, wolves, bears, and other smaller carnivores. Although many caviomorph rodents did survive the onslaught from these digitigrade carnivores, the large-bodied open-landscape species fared poorly and were driven to extinction. As George Gaylord Simpson emphasized, the flatfoot rodents had enjoyed "splendid isolation" in South America for tens of millions of years before the fast killing machines arrived from the north.[12]

This example of rodents gaining release from the evolutionary brakes placed upon body-size increases in plantigrade mammals by the presence of faster predatory digitigrade mammals is not limited to what happened in South America. It is a common phenomenon on many large islands that were colonized by rodents. For example, Madagascar has a giant jumping rat (*Hypogeomys antimena*) that weighs about 1.2 kg (2.6 lb). Not surprisingly, it is highly endangered and clings to life in a tiny geographical space of 20 km^2 (12.4 mi^2) between the Tomitsy and Tsiribihina Rivers north of Morondava in the western dry deciduous forests. On the island of Flores in Indonesia can be found the Flores giant rat (*Papagomys armandvillei*), which also weighs about 1.2 kg (2.6 lb). It shared the island with several related species that were slightly smaller and were driven to extinction by humans. In Southeast Asia giant rats have occurred or do occur on the islands of Timor and Sulawesi and on the Philippine islands. They occurred on islands in the Mediterranean and the Caribbean and on the Canary Islands.

What we are talking about here is the tendency toward gigantism (also referred to as giantism) when small animals colonize predator-free islands. It is called the Island Rule or Foster's Rule.[13] Gigantism does not occur in rodents only. It also occurs in marsupials and insectivores. Of course the opposite of gigantism is dwarfism or nanism, and this occurs on islands in humans, carnivores, and artiodactyls.

The largest lipotyphlan insectivore that ever existed, *Deinogalerix*, lived on the island of Gargano in Italy. It weighed 5 kg (11 lb) and was a gymnure (moonrat) closely related to the hedgehogs. There are eight species of gymnure still living in Southeast Asia, and they barely weigh 1 kg (2.2 lb), which is still very large for an insectivore.

When some plantigrade mammals did manage to escape the constraint of evolving toward larger body sizes applied to them by the existence of larger faster digitigrade carnivores, they turned out to be quite special. This is the same concept, in principal and in part, as the Island Rule's tendency toward gigantism. I say in part, because the Lagomorpha, the rabbits and the hares, provide an example of this release that did not occur on islands; it occurred on the continents.

The lagomorphs have been on earth for a very long time, longer than the rodents or the primates. They appeared very soon after the demise of the dinosaurs, about 66 million years ago. For their size they are the fastest mammals in the world.[14] The European hare, *Lepus europaeus*, for example, weighs 4 kg (8.8 lb) and can run at a top speed of 70 km/h (43 mi/h), much faster than a plantigrade mammal of the same size and faster even than a similar-sized digitigrade carnivore. Lagomorphs were pre-adapted for global cooling—the change from forests to open landscapes—and the appearance of digitigrade carnivores; they have been fast for at least 55 million years.

Gomphos elkema is the extinct ancestor of the rabbits, hares, and pikas (Figure 16.2). The first fossil that was described was found in Mongolia. *Gomphos* had a hind limb almost identical to that of a modern hare. The

10 cm

Figure 16.2. A reconstruction of the 55-million-year-old *Gomphos elkema*, the ancestor of the rabbits, hares, and pikas. The long length of the hind foot indicates that this was one of the fastest runners of the Early Cenozoic. (Image by and courtesy of PanZareta, http://panzareta .deviantart.com)

metatarsals in the foot were highly elongated, indicating that this animal could run fast. When it walked, though, it would have placed its full foot on the ground, just as plantigrade mammals do. So, whereas we'd classify lagomorphs as digitigrade, they also display plantigrade characteristics.

When the digitigrade carnivores emerged during the Eocene, the lagomorphs were ready for them. They had already made the flatfoot break. Other features evolved in them to make them lighter and faster. Their bones became lighter and less dense than those of other mammals, and they had enormous eyes, the hallmark of fast runners. So ancestral lagomorphs were running fast in tropical forests. This is an interesting observation, because *Gomphos* was not the only oddity to have belted it through the forest understory at a time when other mammals still had flat feet. Elephant shrews were, and still are, extremely fast runners, and they also inhabited the Eocene forests.[15]

I have argued that the upper body-size constraint placed on plantigrade mammals is the prime determinant of the shape of the body-size distribution of the mammals on all continents except Australia. Australia does not conform simply because a large assemblage of marsupial predators never evolved there. The evolutionary process squashed the plantigrade mammals into their 2-g to 1-kg (0.07- to 35.3-lb) bell curve because the existence of larger carnivores constrained their capacity to evolve to bigger sizes. The biggest body sizes attained by mammals occur in herbivores, although the average body size of herbivores hovers around the Demment and Van Soest optimum. The digitigrade carnivores fit in between the two distributions, covering a body-size range that allows them to catch and eat small as well as large mammals—in the latter case, with group cooperation. There are exceptions to this basic story, but they are idiosyncratic arguments, and we do not need to get involved with them.

So why is the evolution of body size the secret to understanding why some mammals are hot and others are not? Again, it comes down to that upper body size limit of the plantigrade mammals, the body size at which the upper limits of plantigrade mammal body sizes overlap with those of the smallest digitigrade mammals. This overlap corresponds with an interesting pattern of metabolic rate in mammals. By removing the effects of body size, we can compare metabolic rates among mammals of different sizes. In essence, we can calculate a best-fit line for all data on metabolic rates available for mammals of all sizes and then calculate the residuals.[8] For each species, the residual is the difference between what this best-fit line would predict for an animal of its size and the actual observation. So residuals that fall below the line are negative and indicate lower than average metabolic rates and vice versa. When these residual metabolic rates are again plotted against body size, the pattern looks like a bow tie. Although I showed this in my 2000 publication, I used a later, larger dataset to illustrate the pattern (Figure 16.3).[16] Some of the smallest and the largest mammals deviate the most from the line, whereas mammals in the middle of the logarithmically transformed body-size distribution show the least deviation from the best-fit line. The minimum diversity in metabolic rate occurs exactly at the overlap

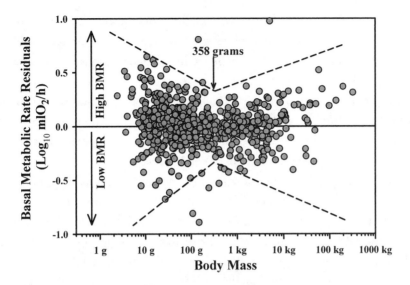

Fig. 16.3. The bow-tie plot of basal metabolic rate (BMR) residuals of 585 mammal species as a function of body mass. Both axes are logarithmically transformed. A majority of the mammals in the data set are smaller than 385 g (13.6 oz), showing the greatest positive and negative deviations from the horizontal line, illustrating a greater variation in metabolic adaptation to their environments compared with larger mammals. The dashed, angled lines are fitted by eye and embrace between them more than 98 percent of the data. (Data from White, C. R., Blackburn, T. M., and Seymour, R. S., "Phylogenetically Informed Analysis of the Allometry of Mammalian Basal Metabolic Rate Supports Neither Geometric nor Quarter-Power Scaling," *Evolution* 63 [2009]: 2658–2667, and reanalyzed)

between plantigrade and digitigrade species at a body size that I have called the "constrained body mass." It is the intercept between the small mammals and the large mammals. For convenience let's say that it occurs at 1 kg (2.2 lb) (the exact calculated value is 385 g, 13.6 oz).

Let's also discuss the small mammals that, for clarity, we consider to be plantigrade mammals smaller than 1 kg (2.2 lb), which include the non-flying orders: Rodentia, Lipotyphla, Scandentia, and some primates. I have excluded bats in this discussion because we are going to talk about migration and territory size, and in terms of the cost of getting around we can't match flying with walking or running.

Being constrained to a small size means being constrained to a small territory. A small territory would occur in a forest, woodland, savannah, desert, or a river. In whichever habitat the small mammal evolved, it would have displayed specializations for exploiting those habitats because it did not have the option of moving to other habitats during poor seasons. For example, in forests these specializations would have included skeletal adaptations for climbing, especially in the hands and feet. Not many mammals can come down a tree trunk head first, as a squirrel does. It has a highly specialized hind ankle that allows its foot to rotate and point backward. It is scansorial (a tree dweller). Small desert mammals need to cover the greatest distances every night compared to other nocturnal mammals in other habitats, because their food is scarce and spread out. What evolved in many separate desert rodent families, such as the jerboas (Dipodidae) and gerbils (Gerbilidae), is saltatorial locomotion, hopping; like kangaroos, they can cover greater distances at a low energetic cost. These are fast rodents, and they, too, have big eyes. Of course small aquatic mammals need to swim efficiently, and most have webbing between their toes.

But consider now the energetic problems of being stuck in one territory. What if there is a long drought, which is, of course, a common feature of deserts? What if the monsoon rains collapse, and there are much-reduced summer rains for that year? What if there is an exceptional cold snap, and the temperature plummets to minus 40°C (−40°F)?

If resource availability is unpredictable in time and space interseasonally, irrespective of whether the habitat is tropical, semi-tropical, desert, Mediterranean, temperate, or tundra, small mammals cannot be hot; they cannot sustain a high metabolic rate.[17] These animals will have the lowest mammalian metabolic rates and body temperatures. They will be basoendotherms.

If resource availability is predictable seasonally, again irrespective of latitude or habitat, metabolic rates can be expected to be normal in small mammals. These animals will probably be mesoendotherms.

If resource availability is predictable seasonally *and* the latitude is high, that is, the winters are cold and unproductive, small mammals will be hot; they will have high metabolic rates and body temperatures. These small mammals will need the metabolic machinery and capacity to produce a lot of internal heat to prevent freezing to death. These will be the small supraendotherms.

That is why the bow tie flares so much at body masses smaller than 1 kg: small mammals have nowhere to go, so they display much greater variation in metabolic adaptation than do larger mammals.

But how can we explain the variation in metabolic-rate residuals for mammals larger than 1 kg (2.2 lb)? Apart from the climatic and latitudinal effects, I do not know of any factors other than locomotor energetics that can presently explain the diversity of metabolism in mammals larger than 1 kg (2.2 lb).[9] The hottest large mammals are indeed the fastest, and the coolest are indeed the slowest. The pronghorn has attained the pinnacle of endothermy; it is the hottest mammal in existence. Even within the carnivores there are body temperature differences related to how the animals obtain their food. The pinnacle of the pursuit predators, cheetahs, are the hottest. The ambling, sniff-and-scratch carnivores, such as badgers and the ratel (honey badgers), are the coolest carnivores.

For the moment, this is my best explanation of metabolic diversity within the mammals.

How well do these ideas apply to birds? Birds, both big and small, can, and do, migrate long distances to avoid habitats that are thermally stressful and/or seasonally poor in resources. Consequently, birds may not be good models for examining the effects of habitats on endothermy because they can escape habitat effects seasonally. Yet despite this confounding variable, Joe Williams and his laboratory at Ohio State University in Columbus have compared tropical and temperate birds and shown that tropical birds have a slower pace of life than do temperate birds.[18,19] Anna Jimenez from his laboratory concluded that tropical birds "tend to have higher survival, slower growth, lower rates of whole-animal basal metabolic rate and peak metabolic rate, and smaller metabolically active organs compared with temperate birds."[19] So it seems that birds share the same patterns of latitudinal endothermy as mammals. Unfortunately, though, there has been considerably less research on avian endothermy, especially as it pertains to differences in flight form and function.

SEVENTEEN

Cool Sperm

It is the worst nightmare of every cricketer or baseball player—those that bat, that is—getting hit in the testicles by the ball in full view of millions of television viewers on a warm summer's day. A player falls fast, into the fetal position, chin out, clutching his two greatest assets, and rolls around in sheer agony. It is awful to watch, but worse to experience. What a bloody stupid place to store a man's genetic arsenal: in two little sacs that dangle between his legs. If wombats could play cricket or baseball, it would not happen to them, and it would not happen to hedgehogs, elephants, or hairy armadillos either. Many mammals do not have the little dangling sacs, the scrotums that house their testicles. Evolution, it seems, works in funny ways, which cricketers and ballplayers likely do not find very amusing.

In animals that have scrotums there is always a high probability that the testes can be damaged. Damage could compromise sperm production and storage, which would, of course, be disastrous to the animal's fitness potential. But if we believe, as I emphasize repeatedly, that evolution works through trade-offs that balance fitness benefits with costs, there must be a benefit to leaving the highly vulnerable testicles dangling. The explanation had better be damn good, too.

The mysterious benefit of externalized testicles eluded biologists for a long time. But there is absolutely no doubt in my mind that temperature has everything to do with the answer. The scrotum is made up of cremasteric muscles that contract and relax to pull the testicles closer to the body or allow them to dangle farther away from it. The muscles twist and churn and keep the testes on the move. No other muscle in the body is so visibly mobile. I cannot stare at the muscles housing my left testicle and command them to move, as I can command my left middle finger to move. I have no control over the movement of these muscles because they are involuntarily controlled by temperature. When the testicles get too cold, the cremasteric muscles contract and draw the testicles closer to the body to warm them up. Every boy who has swum in cold water, say at 9°C (48°F), will have noticed that his testicles disappear into his abdominal cavity looking for heat. He will have noticed, also, that on a hot day his testicles dangle some distance away from his body in his long, pendant scrotum. Scrotal testicles have a marvelous ability to control their temperature at a level that is lower than that of the body, always. When the testes are allowed to dangle, they can control their own temperature. When they are squashed into tight jeans, they do not dangle, and they cannot control their own temperature.

Mammals are not born with pendant testes. In all mammal embryos, the testes are located on top of the kidneys in the abdominal cavity. This is called the testicond position. In some mammals, like elephants, the testes stay there forever. In others they don't. In the latter, the testes start a migration from the kidneys at puberty and descend into the abdominal cavity through the inguinal canal. If the testes stay in this position, sitting on the body wall of the abdomen, they are said to be in an inguinal position. However, in many mammal lineages the testes make a final migration; they perforate the abdominal body wall and descend into the scrotal sacs. This is the scrotal position. Cricketers and baseball players are scrotal.

The testes in the three positions, testicond, inguinal, and scrotal, have different temperatures. In testicond mammals the testes are at the same temperature as that of the body. In the inguinal position, they are also more or less at the same temperature as that of the body, but at certain times of the year the testes can bulge into the abdominal wall, which may cool them slightly through heat exchange across the abdominal wall. But the testes of scrotal mammals are *always* cooler than the body temperature. Puncturing the abdominal wall, the barrier to the interior of the body, must bear with it enormous benefits compared with the cost of compromising the body wall and exposing the sperm factory to dangers, such as curving cricket balls or baseballs.

Did Schrëwdinger have a scrotum? Probably not. I say this with more than 95 percent certainty after reconstructing the ancestral testis position. My analyses, published in the *Journal of Evolutionary Biology* in 2014, showed that the scrotum appeared, quite suddenly, about 64–65 million years ago, in the ancestral mammal that gave rise to the two *über* mammal clades, the Euarchontoglires, which includes rabbits, rodents, tree shrews, flying lemurs, and the primates, and Boreotheria, which includes insectivores (Lipotyphla), bats, pangolins, carnivores, and ungulates.[1] The Afrotheria, which includes elephants, hyraxes, aardvarks, elephant shrews, golden moles, tenrecs, and dugongs, and the Xenarthra, which includes the armadillos, sloths, and anteaters, branched off at the bottom of the mammal family tree before the *über* clades. These animals do not have scrotums; they are testicond. So the testicond condition is clearly plesiomorphic, which means that it is the ancient, ancestral condition. That is why I am sure that Schrëwdinger did not have a scrotum: it was the ancestor of the Afrotheria and the Xenarthra.

This scenario may prove to be incorrect, though, if it can be shown that the testicond condition in Afrotheria and Xenarthra is not the ancestral state but is, instead, a reversal to testicondy or an inguinal state. Testes and scrotums do not fossilize, so this alternative cannot be proved using the fossil record. But, who knows, China may yet produce another stunning fossil showing the dangling bits.

At the moment there are no data I am aware of that could be employed to argue this alternative, although a new genomic analysis was recently published arguing that testicondy was not the plesiomorphic state in placental mammals.[2] The study claims that the remnants of once functional relaxin/insulin-like family peptide receptor 2 (*RXFP2*) and insulin-like 3 (*INSL3*) genes associated with descent of the testes suggest that the scrotal state was plesiomorphic and that the Mesozoic placental mammal ancestor possessed descended scrotal testes. The Xenarthra and the Afrotheria, it is argued, became secondarily ascrotal, despite the fact that some afrotherians (elephants and hyraxes) and all Xenarthrans, possess functional *RXFP2* and *INSL3* genes. If this is valid, it suggests that somewhere between the origin of the Marsupialia and that of the placental ancestor, a pulse in endothermy occurred that might have led to the evolution of the scrotum. The study curiously did not include the marsupials or the monotremes in the analysis, which would have provided insight into an alternative origin of the scrotum. However, given that the Monotremata are testicond, and assuming that monotreme testicondy does not represent a secondary loss of the scrotum, we can assume that the earliest origin of the scrotum under this scenario would have occurred after the monotreme-marsupial diversification. Thus the potential scrotal ancestor would have occurred somewhere within the Eutriconodonta, Multituberculata, Spalacotheriidae, stem cladotherians, or stem boreosphenidians. With the exception of the dinosaur-eating eutriconodont *Repenomamus*, these were all small, nocturnal mammals, < 5 kg, that occupied a high diversity of ecological niches. None showed any evidence of cursoriality, which, in terms of my hypothesis that the scrotum is closely associated with cursoriality, makes it unclear what the evolutionary selection pressures for the evolution of the scrotum might have been in Mesozoic mammals, in which T_b may have exceeded basoendothermic levels.

How can we figure out why this crazy scrotal innovation should have appeared so suddenly after the great dinosaur barbecue? In humans, the scrotal temperature is around 35°C (95°F) at rest, about 2–3°C (3.6–5.4°F) lower than the body temperature. Why does it need to be lower than the body temperature? Why is keeping sperm cool necessary?

There are quite a few hypotheses floating around, but none of them are meaningfully anchored in the evolutionary sequence of events that led to the mammals. Some are focused on temperature, whereas others do not regard temperature as important at all.

Frey's Galloping Hypothesis is a good example of a hypothesis that does not implicate temperature. It argues that the pressure fluctuations that occur in the abdominal cavity are so great when a mammal gallops that it would impair blood flow to and from the testes, compromise sperm production, and damage sperm already in storage in the epididymis.[3] This hypothesis is hard to test. The scrotum did indeed appear when faster locomotion evolved in mammals, so it cannot be dismissed out of hand. But, as I argue later, an alternative explanation is that the scrotum can be linked to the faster locomotion and higher body temperatures that evolved in cursors, that is, fast runners, without having anything to do with pressure problems at all.

The most recent hypothesis, published in 2009, does involve temperature. Gordon Gallup, Mary Finn, and Becky Sammis from the Department of Psychology at the State University of New York at Albany, proposed the Activation Hypothesis, which argues the following: "As a result of consistent temperature differences between the male and female reproductive tracts . . . we propose that the rise to body temperature that accompanies insemination into the vagina serves as one of several triggers for the activation of sperm."[4]

The Activation Hypothesis is testable, and also refutable, which is why I got interested in the evolution of the scrotum, by mistake, when I started trying to figure out how hot Schrëwdinger might have been. I reconstructed its body temperature to have been about 36.2°C (97.2°F). Given the limitation of my method, we can be 95 percent sure that Schrëwdinger's body temperature lay between 35.7°C and 36.8°C (96.3–98.2°F).[4] The limitations concern ghost lineages in the family tree. Ghost lineages are long branches in the tree for which there is no information, and there is no information because body temperatures do not fossilize.

Reconstructing a feature such as body size using the same method provides much more reliable reconstructions of ancestral states because it is possible to estimate the mass of extinct mammals from the sizes of their teeth or the lengths of their long bones. The data for fossil species can then be added to the family trees of living species. The branches of the added species end at the point in time when they went extinct, so not all branches in the tree end at the present. Nevertheless, the extra information that these extinct body sizes provide vastly improves the quality of the estimation of the ancestral mammal's body size. Adding data for extinct mammals to the mix eliminates the problem of ghost lineages—these data *are* the missing ghost data. Sadly, data for the body temperatures of extinct mammals do not exist and cannot be estimated from fossils. Or can they?

Robert Eagle's team of researchers used the method of measuring iso-topes in fossil dinosaur eggshells to estimate the body temperatures of dino-saurs, and, by following a similar method, they used fossil teeth to estimate the body temperatures of an extinct rhino that inhabited Florida during the Miocene, about 12 million years ago. Their estimate was 36.6°C (97.9°F), exactly the same temperature as an estimate that they got using the same technique for the modern white rhino, *Ceratotherium simum*, from Africa.[5] Unfortunately, though, at the time of this writing, these estimates for the extinct rhino and an estimate of 38.4°C (101.1°F) for the wooly mammoth are the only estimates for extinct mammals that have been made so far. It would be extremely informative if estimates could be made of the earliest ancestors of all of the mammalian orders, or even better, all of the families. This would solve the problem of ghost lineages very nicely. Having said this, though, I have some concerns about the body temperature estimate for the mammoth, which I will tackle shortly.

For the moment, let's go with my body temperature estimate of 36.2°C (97.2°F) for Schrëwdinger. My gut feeling is that it may be an overestimate, but the value suits the argument that I am trying to develop, so I won't tinker with it. If Schrëwdinger had a body temperature of 36°C (96.8°F) (I will drop the extra 0.2°C for convenience), what was the body temperature of its ancestors?

If we add data for the marsupials and the egg-laying monotremes to the mix, I can reconstruct the body temperature at the root of the mammal tree at the point where the monotremes diverged from the rest of the mammal lineage. This ancestral mammal was about 166 million years old, and I es-timate its body temperature to have been about 33°C (91.4°F).[6] This would have been the body temperature of little *Hadrocodium*.

What these analyses predict is that the body temperatures of all mammals ancestral to Schrëwdinger in all probability did not exceed 36°C (96.8°F). I also suspect that, in the placental lineage, the Mesozoic mammal did not have a scrotum; it was testicond. Using this logic, we can conclude that the temperature at which spermatogenesis—sperm production—occurred in early mammals never exceeded 36°C (96.8°F). And indeed, since we also know that the true mammals first appeared 166 million years ago, we can imagine that sperm production and storage never exceeded 36°C (96.8°F) for 100 million years prior to the evolution of Schrëwdinger. This is a long time, and sperm production was obviously highly successful for this period of time, at temperatures below 36°C (96.8°F). If it had not been successful, mammals would not exist today.

Over this period of 100 million years of successful sperm production, I suspect that stabilizing selection occurred. In other words, any tendency for the body temperature to rise above 36°C (96.8°F) during the evolution of the mammals would have been "opposed" by natural selection. The value would have been "pulled back" below 36°C (96.8°F), because higher values would have reduced the fitness of the candidate. A temperature of less than 36°C was a characteristic that worked for spermatogenesis and for sperm storage and maturation, and natural selection preserved this successful temperature optimum. I'd extend the argument further, to the Cenozoic, and suggest that stabilization of the selection of the temperature of spermatogenesis has continued right up until the present. To this day, certainly in mammals, sperm maturation and storage in the epididymis still occurs at temperatures below 36°C (96.8°F) despite the fact that the body temperatures of some mammals approach 40°C (104°F).

If I am right, the Activation Hypothesis is flawed. Males without scrotums have been inseminating females for 100 million years. Both male and female would have had the *same* body temperature, so there would not have been a thermal shock to activate the sperm. Indeed, even in modern testicond mammals, such as elephants, sperm does not undergo a thermal shock during ejaculation.

The Eagle team's estimate of 38.4°C (101.1°F) as the body temperature of the extinct wooly mammoth is a concern and challenges the idea of the optimal temperature for spermatogenesis. Mammoths belong to Elephantidae, to which the modern Indian and African elephant belong. Elephants are testicond. Their body temperature is about 36°C (96.8°F), right there at my postulated upper limit for the temperature of spermatogenesis. It is extremely unlikely that the scrotum would have evolved independently in mammoths. Either the Eagle estimates are too high or my cool sperm ideas are flawed. We need more critical assessment of the isotope techniques that have been employed, as well as more estimates of fossil mammal body temperatures before this problem can be resolved.

Why should a temperature above 36°C (96.8°F) compromise sperm production, maturation, storage, and performance, and hence reduce fitness? This is an obvious question without an obvious answer.

The best explanation, I think, concerns potential mutations on the male Y chromosome during sperm maturation and storage. Robert Short from the Royal Women's Hospital in Melbourne, Australia, has argued that the testes are a "hot spot" for germ cell mutations by free radicals.[7] Sperm are not simply produced in the testes in huge numbers to wait their turn to be propelled out of the penis during ejaculation. Actually, they first undergo a complicated sequence of maturation events during a period of storage in the epididymis, and these events consume a lot of energy. The maturation prepares the sperm for its "swim" through the uterus to the fallopian tubes of the female, where it will meet, hopefully, the descending egg of the female about two-thirds of the way up the tube. Of course it has to compete with millions of other sperm for this privilege. Maturation not only kick-starts the mobility of the sperm; it also arms it with penetrating power, a bunch of proteins up front that will allow it to punch a hole through the outer membrane layers of the egg—the *zona pellucida*—to allow fertilization to take place. But the maturation process is vulnerable to temperature.

Arrhenius effects—the kinetics of molecular motion—will increase the rate at which reactions occur, and the number of reactions that occur, during sperm production and maturation. All biochemical reactions that involve oxygen bear bad tidings in quantities wholly proportional to temperature. These are the nasty metabolites that are the byproduct of respiration—free radicals. Free radicals are oxidants, and they cause mutations. A sperm cell contains not only the gear to swim and knock a hole through to the egg; it also carries the male's genetic arsenal, the DNA code for that male, which will be combined with the DNA from the female to make up the full complement of the DNA of the new offspring. The male's DNA, though, is vulnerable to the free radicals. It can be distorted and changed, its molecules swapped or mutated. Mutations are the source of new variations in populations, which is what natural selection feeds on, and thus they are not always bad. They can be selected for in the evolution of new innovations. But what is critical about mutations is the *rate* at which they occur. Temperature-induced pulses in mutations might not be favorable, because they tend to drive evolution at a pace faster than it usually occurs when the rate of mutations is not temperature dependent. One hundred million years of spermatogenesis at temperatures below $36°C$ ($96.8°F$) have presumably minimized the rate of mutations during sperm storage. Stabilizing selection has been the guardian of spermatogenesis, ensuring that high temperatures do not drive the pace of evolution too fast.

The testis produces 176 million sperm cells per day. It has a massive metabolic rate. It uses a lot of oxygen and hence produces a constant number of free radicals. The rate at which these are produced increases dramatically with increased temperatures during the process of storage and maturation in the epididymis.

Unlike in birds, mammalian sperm maturation and storage in the *cauda epididymidis* is a specialized function controlled by male hormones at a lower body temperature to ensure that there will be successful penetration of the unique mammalian *zona pellucida*.[8] For birds, though, mutations during sperm storage at high temperatures do not seem to be a problem. Male birds are homogametic (ZZ) compared with heterogametic (XY) male mammals, so their sex chromosomes can undergo recombinant repair in the advent of mutation. Birds do not have scrotums or any other form of testis externalization, and gram for gram they have higher body temperatures than mammals. All birds have body temperatures higher than the $36°C$ ($96.8°F$) upper limit of spermatogenesis. Why do their Z chromosomes not mutate? Robert Short provides another explanation: "In birds, it is the male that is the homogametic sex (ZZ), and the female is heterogametic (ZW), so they are spared the problem of exposing the heterogametic sex chromosome to the mutagenic environment of a testis, which may reduce the need for testicular cooling. And, in any case, perhaps birds are more able to cope with a high mutation rate because of their oviparity. Because most mutations are likely to be deleterious, then laying an infertile egg in a clutch or having a chick die in the nest is not a great waste of reproductive effort."[7]

I am not sure why this explanation should not apply to neonate mammals as well.

Schrëwdinger gave rise to the placental mammals of the Cenozoic, the mammals that we know today. Many of these mammals have body temperatures way above 36°C (96.8°F), and yet they produce and store viable sperm. How do they do it?

On land there is one way only, by dangling the testes away from the hot body. Whales and dolphins have abdominal testes, but their problem is solved by a countercurrent blood flow system that cools the testes with blood that has been cooled in the fins.[9]

Who are the scrotal mammals, the testes danglers, apart from cricketers and baseball players?

The easy answer to this question is that they are the hottest mammals. The hottest mammals are the rabbits, hares, ungulates, and carnivores. They all have scrotums. What the nonhumans all have in common is that they can run fast. They are cursors.

As I mentioned in earlier chapters, Andrew Clarke from the British Antarctic Survey, Cambridge, and Hans-Otto Pörtner from the Alfred Wegener Institute in Bremerhaven have argued that heat is essential for producing the ATP needed for muscle power.[10] For sustained aerobic respiration during running, ATP must be generated rapidly to fuel the energy demands of the muscles. These authors have emphasized very clearly that ATP generation by the mitochondria is strongly dependent on temperature. Warm muscles are necessary to sustain fast locomotion.

The release of the mammals from the dark of the Cretaceous threw open unparalleled opportunity for them to claim the multitude of habitats that the dinosaurs vacated at the end of the Cretaceous. These niches were everywhere, in the seas and inland waters, on land, and in the air. In the air, competition was a little stiffer because not all of the dinosaurs died; one lineage survived — the birds. On the ground, mammals could get bigger, and some could again become active during the daytime. And indeed, this is exactly what happened, with astonishing evolutionary speed within the first several hundred thousand years of the Cretaceous-Cenozoic boundary.

The pace of evolutionary change went berserk during the Early Cenozoic. From Schrëwdinger two new mammals evolved, one that gave rise to the Afrotheria and the South American Xenarthra and another that gave rise to all the others mammals, including us. The split happened 66 million years ago. Further radiations within these two descendants of Schrëwdinger followed rapidly, giving rise to every placental mammal order that exists today, and again these splits happened very soon after the meteorite impact, within half a million years. This was a truly remarkable explosion of new mammalian forms — from Schrëwdinger to the ancestors of the Afrotheria, Xenarthra, Lagomorpha, Rodentia, Scandentia, Dermoptera, Primates, Pholidota, Chiroptera, Lipotyphla, Carnivora, Artiodactyla, and Perissodactyla, all within a mere 500,000 years, a blink of the eye in evolutionary time.

As descendants of Schrëwdinger, the thirteen new mammal orders were good to go, and to avoid competition with one another what evolved in them was a diversity of form and function, innovations necessary for them to fit into the new freed-up niches. Evolution during the Cenozoic was very heavily piloted by global cooling, which commenced about 30 to 40 million years ago.

Evolution had come full circle. The ancestors of the pelycosaurs, stem amphibians such as *Diadectes*, were the first in the lineage that led to mammals to start eating plants. But herbivory became redundant during the long Mesozoic stint in the dark because the nocturnal mammals became small and focused on eating bugs, snails, fish, amphibians, and small reptiles. Some even ate dinosaurs. Small body size and reliance on the energy from cellulose fermentation *alone*, as I have emphasized on several occasions, are simply incompatible. The fitness benefits that drove the miniaturization of the mammals during the Triassic and the Jurassic are not clear, but we can be fairly certain that they had nothing to do with eating plants. Herbivory was abandoned in all but some multituberculates that went extinct in the Cenozoic. But the capacity to digest cellulose returned during the Cenozoic for a second time in the mammalian lineage to allow mammals to again dominate the use of the earth's food resources.

For herbivores, the ability to exploit grasses required the capacity to follow the food resources seasonally and to avoid being eaten by predators in the process. Both of these fundamental avenues to fitness demanded two things—a new set of legs, and heat. Schrëwdinger's heels-on-the-ground plantigrade feet and short little legs were not going to enable its escaping from fast predators or walking long distances to find food in tall grass. I'm not saying that, because Schrëwdinger's foot was ancestral, it was therefore useless. It was, and still is, the perfect little foot for living as an insectivore or as a primate, a mouse, or a hedgehog. Schrëwdinger did not need to run fast to escape predators or find food. She could climb trees, dig holes in the ground, and scratch around in the leaf litter looking for grubs and could groom herself with her hands and feet. Her plantigrade feet were perfect for a small, nocturnal insectivore. And indeed today most mammals remain plantigrade and nocturnal.

Open landscapes made it easier for carnivores to see their prey, but they needed more speed to catch the herbivores. Those that evolved to eat the new open-landscape herbivores needed to become pursuit predators. The herbivores that they desired to eat also benefited from more speed to avoid being caught, and they also needed greater awareness, bigger eyes, and enhanced smelling and hearing abilities. In the early times after the dinosaur extinctions, when the earth was still heavily forested, carnivores caught their prey in the forests by ambush, which did not require speed. Ambush hunting was not lost when the landscapes opened up—far from it. Many carnivores today, such as lions and leopards, still rely on ambush hunting in addition to sheer burst speed and pursuit hunting. But open-habitat ambush hunting required greater bursts of speed and acceleration.

As a generalization, we can work on the simple premise that the first way to increase speed is to increase stride length, the distance between one footfall and the next. All marathon runners know this, and they also know that overstriding causes injury, because the leg hits the ground when it is straight and the shock of the impact is carried to the joints. If the leg is slightly bent at the knee when the foot hits the ground, the leg acts as a shock absorber, and there is less of a chance that an injury will occur. So stride length is a rather nonnegotiable variable for runners — it is a fixed length for each individual. My stride length is 990 mm (39 in).

The second way to increase speed is by increasing cadence, that is, the number of steps per minute. All marathon runners know this, too. The rule-of-thumb target for marathon runners is 180 steps per minute. My average cadence is 166 steps per minute. Cadence is not set. It can, and is, altered to change running speed. It is the accelerator pedal. My pedal seems to jam at 166 steps per minute.

Here I am drawing analogies with human locomotion. However, humans are oddballs when it comes to locomotion and are awful models for comparison. We run on two legs. No other mammal does this routinely. We also have flat feet; our heels hit the ground. No fast mammalian cursors are heel strikers. Nevertheless, the concepts of stride length and cadence apply to all mammals, and familiarity with our own running foibles helps to illuminate the concept.

The third trick that evolution exploited to increase speed was selection for longer legs, which automatically increases stride length. The change in the legs of large carnivores and herbivores during the Cenozoic is one of the spectacular showpieces of natural selection and evolution. It was an epic bone show popularized best by George Gaylord Simpson, America's most esteemed paleontologist during the first half of the twentieth century. Simpson used horses to illustrate his show and, in his book *Horses*, he graphically showed the changes in the limbs that occurred from those of the ancestral browsing horse to those of the modern grazing horse.[11]

Before we continue, I need to take you on a short anatomy refresher. Consider your legs. The single big bone under your quadriceps and hamstring is your femur, the biggest of the long bones in the body. It is connected to your hip with a ball and socket joint. The femur is the site of the attachment of many muscles used for locomotion and to keep the body upright. Below the knee are two bones that lie side by side, sort of, the tibia and the fibula. The tibia is the major weight-bearing bone in the limb, whereas the fibula is involved in the rotation and twisting of the limb. Attached to these two bones is the foot. There are three groups of foot bones: the tarsals, which are the ankle bones and include the heel bone (calcaneum); the metatarsals, which are the main bones of the foot; and the phalanges, the bones of the toes.

To lengthen the leg, natural selection targeted the metatarsals, the bones in the middle of the foot. The feet of cursors got longer in the middle. What you perceive to be an antelope's backward-pointing "knee" is actually its heel, highly elevated off the ground. The other long bones in the leg, the femur, the tibia, and the fibula, could not be targeted for proportional change because increasing their length relative to that of other long bones would have compromised the animal's structural strength and weight-bearing capacity. The long bones could potentially snap in half if they got too long. Look at what has happened to thoroughbred racehorses. Humans have, of course, taken over the role of natural selection and have selected artificially for longer legs so that thoroughbreds can run fast on nice smooth tracks. But when these horses stumble, even slightly, they can break their legs. It is estimated that three thoroughbreds are shot every day during training and racing in America after breaking their legs. None of the living wild horse species and zebras have legs anywhere near the same length as racing thoroughbreds, and they do not break their legs when they stumble. They also do not run on nice smooth tracks.

Because metatarsal lengthening became the focus of selection in the Cenozoic, the length of the metatarsals relative to the other foot bones, which did not change much in length, has become a good measure of how well limbs are adapted for speed. The bone that was chosen for comparison was the femur, because it shows the least proportional modification associated with increasing cursoriality. The ratio is called the metatarsal:femur ratio (MT:F). The average human femur length is about 480 mm (18.9 in) and the metatarsal length about 70 mm (2.8 in), so the average human MT:F would be 70/480 = 0.15. This value is typical for plantigrade mammals like rodents, primates, insectivores, and you and me.

The MT:F ratio is a useful index for identifying cursors among the extinct mammals. It is also useful for identifying trends that occurred throughout the Cenozoic. It is quick to measure and easy to compute simply by measuring the long bones and foot bones of fossil mammals. It provides an immediate indicator of whether or not an animal was a cursor. The same ratios can be calculated for the forelimbs from the ratio of the metacarpals to the humerus.

Christine Janis, whose work we discussed in Chapter 2 concerning the development of costal breathing, and Patricia Wilhelm analyzed the MT:F ratios of carnivores and herbivores throughout the Cenozoic. They showed that, by the Late Eocene, the MT:F of dawn horses was about 0.65 and reached values of around 0.95 by the mid-Miocene.[12] Early camels had the highest MT:F ratios, often approaching or even exceeding 1. Many modern herbivores have MT:F ratios in excess of 1, which means that their metatarsal foot bones are longer than their femurs. Gazelles have MT:F values of about 1.14, whereas the fastest of all herbivores, the North American pronghorn, *Antilocapra americana*, has a value of exactly 1.

The pronghorn is the fastest land mammal in the western hemisphere, which is somewhat puzzling because there is currently no predator in America that can outrun it. It runs as fast as 100 km/h (62 mi/h), but which animal chased it to this evolutionary pinnacle? The predators that did chase the pronghorn, cheetahs, lions, and hyenas, are extinct. They were partially the victims of humans, driven to extinction during the past 20,000 years. Many authorities argue that the ravages of the ice ages were also partially responsible for the demise of the great American megafauna, such as the ground sloths, rhinos, and horses. Although many species of *Antilocapra* pronghorns joined the long list of mammals that went extinct on the American prairies, *Antilocapra americana* survived and flourished. It survived the killing efforts of humans as well as the dynamic perturbations of climate change and the ice ages.

In his essay "Ghosts Chasing Ghosts: Pronghorn and the Long Shadow of Evolution," Michael Branch, professor of literature and the environment at the University of Nevada, Reno, pours praise on this remarkable herbivore: "Survived may be too grim a word for so beautiful a product of so beautiful a process. Say instead that pronghorn have been turned on the lathe of evolution for twenty million years, sculpted by predator and place, fired in evolution's prairie and desert furnace. Seen in the light of its evolutionary history, pronghorn is not a thing but rather an outcome—one as inevitable as it was unlikely."[13]

The evolutionary bent toward increasing running speed in herbivores did not stop at lengthening the metatarsals. More was needed—much more. To increase cadence, the lower part of the limb needed to be light and simple. Herbivores do not need to twist or rotate their hind legs because the shift to fast running also demanded a commitment to parasagittal locomotion, that is, movement of the legs forward and backward without any sideways movements. The outcome was that the fibula started to become redundant; it got smaller, and in some herbivores, such as modern horses, it merged with the tibia and disappeared altogether.

During the Early Eocene, 50 million years ago, *Hyracotherium*, an early forest-dwelling, dog-sized ancestor of the horses, stood about 200 mm (7.9 in) high at the shoulder (shown in Figure A.3). Reconstructions have given *Hyracotherium* a patterned, camouflaged coat, suggesting that hiding was important to its forest survival. It was a browser. Nevertheless, *Hyracotherium* was probably one of the fastest runners of its day. The fibula of *Hyracotherium* had fused with the tibia to form a single bone below the femur in the hind limb. This reduction in the number of long bones continued in the later dawn horses and still exists today in modern horses and zebras.

Hyracotherium was also the earliest herbivore to show the next major innovation toward limb lightening and simplification: the reduction in the number of fingers and toes. It had four toes on its front feet and three in its hind feet. Each toe had a hoof, and the metatarsals were already quite long. By the Late Eocene, 35 million years ago, *Mesohippus*, a descendant dawn horse, had lost a further toe. *Mesohippus* was slightly bigger than *Hyracotherium*, about 600 mm (23.6 in) at the shoulder, and was also a forest browser. By 15 million years ago, the number of toes had been reduced to one on both the front foot and the hind foot in the dawn horse *Pliohippus*. *Pliohippus* was beginning to resemble horses as we know them today in size and shape. It was a grazer on open landscapes. It had a single, hoofed foot on each leg. The hoof is a modified fingernail, so *Pliohippus* ran on the tips of its toenails and fingernails. This form of locomotion is called unguligrade because it occurs in all ungulates.

The fibula, four out of five metatarsals, and lots of phalanges were no longer functional, so the muscles that were involved in limb rotation and foot movement became markedly reduced, which also lightened the lower limb, facilitating a faster cadence. The leg muscles now became bunched up around the femur, and the lower leg became long and skinny. These were prime innovations for running fast and for traveling great distances, but they came with one very big sacrifice: ungulates no longer had functional fingers and toes. They could no longer climb trees. They could not dig or groom themselves. They could not pick up food and hold it to their mouths.

If we see the loss of fingers and toes as potential fitness costs, there must have been a massive counterbalancing fitness benefit. The benefit was speed, which allowed these animals to outrun predators, and endurance, the capacity for long-distance cheap transport and the ability to transfer one kilogram of meat for one kilometer at the lowest cost. Birds can do it more cheaply, but only because they fly faster than mammals can run.

So far we have ignored the biggest cost of all, the real cost, in joules and degrees centigrade, of fast running. Long, thin legs with single toes and fingers provided an excellent chassis for speed, but to attain the speed, the chassis needed an engine. The chassis needed hot, powerful muscles.

The bearer of the hot body and fast limbs needed to be ready to run within a split second, day or night, hot or cold. Predators attack when they perceive weakness in vigilance. Remember, too, that in addition to getting faster, both the ungulates and the carnivores also got bigger throughout the Cenozoic for reasons that I discussed in the previous chapter. Big ungulates can't dig their own holes in the ground and hence cannot hide underground during their daily resting phase as so many small mammals do. Ungulates cannot switch off for one second. They cannot hibernate, and they cannot use daily torpor. They need to be in a state of constant readiness unless it is absolutely certain that no predators exist.

The pronghorn has a brain temperature of 39.9°C (102.7°F) and an abdominal temperature of 38.8°C (101.8°F).[14] I checked to see which ungulates might have temperatures higher than this and, to my disappointment, there are two: the domestic sheep and goat. But here we need to interpret the data with caution. Wherever humans have had a hand in selecting the characteristics of domestic stock, both the metabolic rate and the body temperature go up. It probably has to do with artificial selection for faster growth rates to increase production, something that might not occur under natural selection in the wild. Apart from these domestic hotties, the next hottest ungulate is the reindeer, *Rangifer tarandus*, with a body temperature of 39.2°C (102.6°F). It is reassuring in terms of the ideas that I develop here that highly cursorial nondomestic ungulates do indeed exhibit some of the highest temperatures in the mammalian body.

So far we have discussed the herbivore line, the artiodactyl (even-toed) and perissodactyl (odd-toed) ungulates. But it was the carnivores, of course, that pushed the ungulates to the pinnacle of mammalian heat. What about their legs? How hot did they become?

The modern carnivores evolved from a group of Paleocene mammals called the creodonts. Creodonts were plantigrade and could probably not run very fast. However, in the first true carnivores of the lineages of the cats and dogs that exist today, the foot bones began to get longer, and the heel became elevated off the ground, as it did in the ungulates. The carnivores started to walk on their fingers and toes, on padded paws that ended in claws, not hoofs. Finger and toe walking is called digitigrade locomotion, walking on the digits of the fingers and toes. Of course this midfoot length-ening of the front and back legs increased stride length and therefore speed. But this is where it ended. In carnivores the fibula did not disappear, and the number of fingers and toes was reduced, at most, by one. There is a very good reason for this: carnivores needed their claws to catch their prey and could not sacrifice the functionality of fingers and toes as the ungulates did. Consequently, the lengthening of the metatarsals could not equal the measures that were obtained by the ungulates during the Cenozoic preda-tor-prey arms race. The MT:F was mostly less than 0.4, slightly more than double that of yours.

Theoretically, carnivores should not be able to run faster than their prey. Yet the fastest land animal is a carnivore, the cheetah. Cheetahs can attain speeds in excess of 110 km/h (68 mi/h). To find a possible explanation for this outlier, we need to look at all of the options for running speed and try to figure out how the cheetah can run so fast. The MT:F of the cheetah is

0.44, which is not the highest, even for a felid cat. It therefore does not have legs that are any longer than those of other cats.

The secret of the cheetah's speed is cadence and stride length. This critically endangered species is able to arch its spine with each stroke, which has the effect of lengthening the stride length. The stride length has been measured at 7.6 m (25 ft), which is eight times more than mine! When it starts running, it hits this stride three times a second, so that's 22.8 m (72 ft) per second or 82 km/h (51 mi/h). That's zero to 82 km/h (51 mi/h) in one second, faster than any Formula One car.

The cheetah has the highest body temperature of all carnivores, which neatly supports the link between speed and heat. But cheetahs pay an enormous price for their high speed. With the need to be lightweight, they are puny animals and cannot compete with larger carnivores when defending their kills. They also cannot run fast for very long, unlike the pronghorn and many other ungulates.

North of Windhoek, the capital of Namibia, on the edge of the great Kalahari Desert, is where Chewbaaka lived. He's dead now; he died on 15 April 2011. Chewbaaka was the world's most famous cheetah. He was the first cheetah ambassador of the Cheetah Conservation Fund.

On several occasions I went with Laurie Marker, the director of the fund, to take Chewbaaka for a run. He used to run on an open field, a piece of savannah that had been cleared to plant tomatoes without irrigation. The piece of open land is so big that you cannot see the end of it, like the sea. The farmer shot himself after several unsuccessful crops, and the farm was bought by Laurie to act as a base for cheetah research and conservation in Namibia.

Chewbaaka chased a lure. It was truly magnificent to see the planet's most specialized speedster doing what no other mammal can do. His acceleration was breathtaking to watch, a blur to full speed. But what most impressed me was how long it took for Chewbaaka to recover from his sprints. After one, he lay under an acacia tree panting for about fifteen minutes, repaying his oxygen debt.

To understand oxygen debt, we need to understand the diversity of locomotor capacities that have evolved in animals. Al Bennett is a world authority on this diversity. As he has said: "The animal kingdom consists of organisms of very diverse activity capacities, maximal performance levels of which an animal is capable. Some animals are completely sessile, some are slow as snails, some are fleet but tire quickly, and some possess seemingly endless stamina. Physiologists, behaviourists, ecologists and natural historians have been interested in quantifying these activity capacities, because animal athletic abilities are of intrinsic interest and have functional and ecological implications."[15]

The category of animals that has interested us here is those that are "fleet but tire quickly." When an animal bursts into a high-speed form of locomotion—such as I have described for Chewbaaka or is seen in a lizard darting after a grasshopper, a crocodile launching itself at a wildebeest, or the fastest man ever, Usain Bolt, doing the 100-meter sprint—the muscles undergo an explosive increase in metabolic rate that generates a massive surge of ADP and demand ADP's immediate replenishment. In the first second, the stores of ATP in the body are depleted. Thereafter the ADP is replenished both anaerobically and aerobically. If the exercise remains very intense, as occurred in Chewbaaka's sprints, within the first ten seconds or so the ATP is provided by creatine phosphate, which donates a phosphate anaerobically to ADP to form ATP in the absence of oxygen. Usain Bolt can theoretically complete his sub-10-second dash without ever having to take a breath. However, creatine phosphate becomes rapidly depleted and is regenerated only when an animal is at rest again. If the intensity of the exercise remains high, ATP is again generated anaerobically following glycolysis—the breakdown of glucose or its stored form, glycogen—into pyruvate, which is then converted into lactic acid (lactate) with the production of ATP. The lactate builds up in the muscles and is thought to cause muscle fatigue. The last source of ATP is generated through aerobic metabolism, again following glycolysis, but in this case the normal respiratory pathways are followed: pyruvate enters the Krebs cycle and is stripped of its electrons, which then move down the electron transport chain in the inner membrane of the muscle's mitochondria, generating a massive proton-motive force, which is the driving force for the conversion of the phosphate-depleted ADP back to ATP. The protons are snapped up by oxygen to form water. That is why we need oxygen to remove the protons once they have crossed the mitochondrial membrane. No oxygen, no aerobic metabolism. Aerobic metabolism, though, is the routine daily form of respiration because you and I don't spend the whole day sprinting.

During these massive surges of activity, the respiratory system can supply only so much oxygen per unit of time to meet the stupendous demands of the muscles. The amount of energy that Usain Bolt uses during his dashes cannot even remotely be met by the oxygen in the breaths of air that he may or may not have taken in the ten seconds or so that he took to cover 100 meters. Yet he does not collapse in a heap 50 meters down the track when his aerobic energy supply outstrips the massive energy demands of his muscles. He finishes his run, probably even faster in the final 50 meters, propelled with the assistance of anaerobic metabolism. No oxygen is used during anaerobic metabolism. So has Bolt essentially cheated in a physiological sense by not using oxygen to fuel his whole dash? No, he repays the debt after the run. In short, during his recovery, the shortfall in oxygen that was not used is inhaled to repay debts: the reestablishment of creatine phosphate and ATP, reoxygenation of the hemoglobin in the blood and the myoglobin in the muscle, and resynthesis of glycogen. This amount of oxygen is called the oxygen debt.

But there is another reason that Chewbaaka sat there panting under the tree recovering from his sprints. It is also related to anaerobic metabolism, but it does not involve either limited oxygen supply or lactic acid buildup. It is about acidity: acidosis. Acidosis is the rapid increase in the acidity of the blood during anaerobic metabolism. The increase in acidity changes the acid-base balance in the blood, which is a physiological feature of endotherms that *must* be maintained at a constant level. For many decades, students of biochemistry were taught that it is the increase in lactic acid that makes the blood more acidic. Not so. We now know that, despite all of the beautiful graphs that have been published over the decades showing the positive relationship between lactic acid and blood acidity, it is not the lactic acid that decreases the blood's pH. It is a real relationship in that the two parameters are clearly correlated, but, as explained by Robert Robergs and colleagues at the University of New Mexico, it is not a cause-and-effect relationship: "There is no biochemical support for the construct of a lactic acidosis. Metabolic acidosis is caused by an increased reliance on non-mitochondrial ATP turnover. Although muscle or blood lactate accumulation are good indirect indicators of increased proton release and the potential for decreased cellular and blood pH, such relationships should not be interpreted as cause and effect."[16]

The real source of the surge in protons (H^+) that are a measure of acidity is the process of glycolysis itself, that is, the nonmitochondrial biochemical events that precede the conversion of pyruvate into lactic acid. So apart from the oxygen debt that Chewbaaka needed to repay, he also needed some time to restore the acid-base balance of his blood.

I have examined the body temperature data for ungulates and fast carnivores in some detail. Their body temperatures increased gradually over the Cenozoic, from Schrëwdinger's body temperature of 36°C (96.8°F) to the ungulates' of over 39°C (102.2°F), an increase of more than 3°C, and this is a conservative estimate.[6] These high body temperatures are well above the upper limit for spermatogenesis. So do carnivores and ungulates have externalized testes held within a scrotum? They do indeed, except for the rhino, whose testes have secondarily retreated back into its abdominal cavity. The only reason that I can think of for this internalization is that the testes were at great risk of being squashed by the enormous thighs of the rhino. The rhino, consequently, has the lowest body temperature of all of the ungulates. The white rhino *Ceratotherium simum* has a body temperature of 35.2°C (95.4°F), which is below the upper limit for spermatogenesis.

It is not surprising that all of the fast cursors have an endearing look about them; they have big eyes. Cheetahs have the biggest eyes of all cats. Hares and rabbits have eyes bigger than other mammals of the same size. Elephant shrews have enormous, gorgeous eyes.

Amber Heard-Booth and Christopher Kirk of the Department of Anthropology at the University of Texas at Austin have examined the relationship between eye size and running speed.[17] They tested Leuckart's Law, which states that animals that are fast need bigger eyes to enhance their visual acuity and help them avoid collisions with obstacles. A bigger eye casts a larger image on the retina. Indeed, these authors showed that mammals follow Leuckart's Law better than birds. Notwithstanding these differences, there is a link between being hot and running fast; enhanced visual acuity requires more complex and rapid neuronal processing in the brain. This capacity is not free; it has a metabolic cost, yet another slice taken from the hot metabolic pie.

Speed is relative. A cheetah has been clocked at 113 km/h (70.2 mi/h), but how do we fairly compare this speed with that of a mouse? We need to factor in the animal's body size. If we want to compare running speeds between mammals, we need to place handicaps on certain mammals if their sizes alone allow them to run fast. One way to do this is to divide the animal's speed by its body length or body mass. This would give us speed units of body size per unit of time. When you do this, the cheetah is no longer the fastest mammal on earth: the Lagomorpha, the rabbits and the hares, overshadow it.

For their size, the Lagomorpha are the fastest mammals. They are built for speed and, like the cheetahs, are lightweight and relatively puny. Hares have a locomotory mode that is a mixture of plantigrade and digitigrade. When they do slow hops, their heels are on the ground. When they sprint, only their toes make contact with the ground. Their bones are lightweight and thin. In their front legs the radius and ulna are tightly associated and, although not actually fused, function as one bone. In their hind feet, the metacarpals are lengthened; the MT:F attains values of about 0.55. Most hares and rabbits have body temperatures higher than 39°C (102.2°F). They are true supraendotherms, and they have scrotums.

All of the hot mammals that have scrotums and run fast that we have discussed are large mammals. They are larger than 1 kg (2.2 lb). But there is one mammal that is small, about 50 g (1.8 oz), that can run faster than all other small mammals. It is the elephant shrew.[18] These endearing creatures are insanely fast. They are a good example of microcursoriality. Elephant shrews may pose a real threat to my thesis that the fastest mammals need hot muscles; elephant shrews do not have body temperatures anywhere near the 39°C (102.2°F) of ungulates. The elephant shrew can run fast not because of its metabolism but because of its extraordinary long and skinny legs. I'm not sure, but I'd guess that elephant shrews attain fast speeds with lowish body temperatures because their dashes in the wild are short, tens of meters long, and they probably rely on anaerobic metabolism to do this, as does Usain Bolt.

EIGHTEEN

Why We Are Hot

In earlier chapters I opted not to formalize the various hypotheses that have been proposed over the decades to explain the evolution of endothermy because I did not want my writing to be too scientific at the outset. I did not want to scare off nonspecialist readers. But now it is time that we sift among some of the options and make some choices, if we can, or at least tweak and twist some of them.

The model of endothermy that has received the most attention to date is the Aerobic Capacity Model. Let's recall what it claims.

Al Bennett from the University of California at Irvine and John Ruben from Oregon State University proposed the model in 1979 in a paper published in the prestigious journal *Science:* "A principal factor in the evolution of endothermy was the increase in aerobic capacities to support sustained activity. . . . We believe that this increased stamina and sustainable activity were important selective factors from the outset in increasing resting levels of metabolism during the evolution of endothermy. To achieve these higher levels of performance, resting metabolic rate was increased."[1]

One of the central tenets of the Aerobic Capacity Model is that the capacity to attain a high metabolic rate during sustained locomotion is linked to the minimum metabolic rate of an endotherm. For example, mammals with low resting metabolic rates would not be able to achieve the high aerobic capacities or the high metabolic rates during running of mammals with high resting metabolic rates. Bennett and Ruben argued that natural selection can "see" the maximum metabolic rate but not the resting metabolic rate. So if there was selection for the higher maximum metabolic rate required for sustaining aerobic capacity during locomotion, the resting metabolic rate was, let's say, dragged along with it, an indirect consequence of natural selection. The resting metabolic rate, they argued, cannot be "seen" by natural selection.

Many studies have tried to test the model. Some have supported the model, and others have not. The method of science does not accept probabilities of failure of more than 5 percent. Ordinarily, one might suppose, the idea should have been turfed long ago, but it has not been. It has endured to this day because of the widespread recognition that it is not a bad idea. The problem is not with the original idea but with its testability and the variety of methods that have been tried to test it.

The most common approach has been to examine the relationship between resting and maximum metabolic rate in as many species of mammal and bird as possible and, using the modern comparative method, to exclude effects such as body size and relatedness and verify whether the two variables are indeed correlated with each other. Well, they are, and quite well, too. However, this approach does not adequately test the Aerobic Capacity Model because, even if there was a stunningly good match, it does not mean that there is a cause-and-effect relationship. You could measure the arm and leg lengths of a thousand people and also come up with a good relationship between the two measures, but it does not mean that the length of the arm is what determines the length of the leg. As Michael Angilletta and Michael Sears argue: "How can selection for increased aerobic capacity explain the relatively high contribution of the viscera to SMR [standard metabolic rate]? More important, how could selection for greater aerobic capacity be linked to the rate of thermogenesis caused by leaky membranes? The lack of a mechanistic link between the processes that contribute to SMR and those that contribute to sustained aerobic activity has cast doubt on the aerobic capacity model."[2]

Despite a strong between-species relationship, the link between basal metabolic rate and maximum metabolic rate falls apart when these variables are measured within single species—in other words, using intraspecific analyses. David Swanson from the University of South Dakota and his collaborators collected these data in several bird species—juncos, goldfinches, chickadees—and found no relationship between the two variables.[3]

Another approach has been to use selective breeding, most commonly with mice. Young mice are put to the test on a treadmill. Their maximum metabolic rate is measured by pushing up the speed until the metabolic rate hits a ceiling. The metabolic rate of the mice is also measured when they are resting. Many parameters are measured, such as maximum running speed, body temperature, body size, and so on. The trick, then, is to select the outliers—those with the highest metabolic rates, running speeds, body temperatures, and so on—and breed these champions with other unrelated champions. Their offspring are then measured to see whether the maximum and resting metabolic rates are linked and whether these characteristics were indeed passed on to the pups. Then the pups with the highest maximum metabolic rates and running speeds are bred, and the process of artificial selection goes on and on. You need a lot of patience to do these sorts of studies and a lot of mouse food. Yet the outcomes of these studies remain inconclusive; half support the Aerobic Capacity Model, and half do not.

There is an exciting new molecular technique called comparative genomics that is gaining ground rapidly, which involves identifying and analyzing genes that are involved in metabolism, for example, those encoding amino acids in the proteins involved in respiration and oxidative phosphorylation. The base sequences in these genes are compared between parts of the family tree to see when and where bursts of sequence changes occurred that led to new amino acid substitutions. These bursts of molecular and metabolic change can then be linked to new metabolic innovations, most importantly, enhanced metabolic rate capacity.

Lawrence Grossman and his colleagues from Wayne State University, Detroit, have used this technique to argue that genes in the neocortex of the brain that encode for proteins involved in the electron transport chain component of respiration have undergone the greatest substitution rates in apes and humans compared with other primates.[4] They link these changes to an increase in the speed of the chain in its capacity to transport electrons from the Krebs cycle to the site of oxidative phosphorylation on the mitochondrial membrane. A faster rate of electron transport means faster ATP generation, high rates of oxygen consumption, and higher rates of metabolism. The authors provide their explanation as to why endothermy was enhanced in those primate lineages that are "smart": "These molecular evolution changes are likely to be linked to the major phenotypic changes that are associated with anthropoid primates including enlarged neocortex, prolonged fetal (and therefore prenatal brain) development and extended lifespan—because all are supported by adaptations in aerobic energy production. The neocortex is a primary energy consumer, yet it must consume 'clean' energy."[4]

Roberto Nespolo and his colleagues from Universidad Austral de Chile and Pawel Koteja from Jagiellonian University, Krakow, Poland, suggested that it is now timely for comparative genomics to be meshed with methods in quantitative genetics and functional genomics to provide the definitive tests of the Aerobic Capacity Model.[5]

Al Bennett and John Ruben argued that increased body warmth per se was not a focus of selection. To support their stance, Bennett and several colleagues performed a forced-feeding experiment on lizards.[6] For their ectotherm model they chose the savannah monitor lizard (*Varanus exanthematicus*), with a body mass of less than a kilogram (2.2 lb). They injected macerated beef through a tube directly into the lizard's stomach. The experiment relied on earlier observations that the metabolic rate of the internal organs in lizards increases substantially following a large meal as a consequence of tissue growth and the metabolic cost of processing the food. The experiment was conducted at 32°C (89.6°F) and at 35°C (95.0°F), and the metabolic rates of resting lizards increased three- to four-fold after forced feeding. Here is the beginning of their concluding paragraph: "In our experimental animals, increased energetic metabolism equivalent to a several-fold increment in metabolic rate had almost no impact on increasing or stabilizing body temperature. . . . It is difficult to see how natural selection would continue to increase metabolic rate if the only goal were a thermoregulatory condition that was not being established." These authors went on to conclude that "the elevated metabolic rates associated with endothermy in mammals and birds probably evolved initially for other selective reasons; and . . . endothermy and homeothermy may be adaptations that developed later than the evolution of increased metabolic rate. From this viewpoint, the thermoregulatory condition of mammals and birds would be a secondary refinement of an elevated metabolic condition developed by other selective factors."

I agree entirely, as I indicate in my timeline model discussed later. The body temperature of the lizards increased by a paltry 0.5°C (0.9°F)—extremely poor economics—a four-fold energy investment for a 0.5°C increase in body heat. Of course heat was generated by the increased metabolism—it always is when ATP is hydrolyzed—but the heat was lost so rapidly from the lizards that it barely raised their body temperatures. Lizards do not have insulation in the form of fur or feathers to trap the heat, so it was hardly reflected in their body temperatures.

Roger Seymour, whom I have mentioned in earlier chapters, compared the muscle power of today's ectothermic crocodiles with that of endotherms of the same size after heating up the crocs to the same body temperature as the endotherms. He lent support to the Aerobic Capacity Model by showing that the muscle power of a 200-kg (441-lb) crocodile was a mere 14 percent that of similar-sized endotherms and emphasized that muscle temperature alone cannot confer upon ectotherms a higher aerobic capacity.[7] What crocodiles needed, in addition, to attain equivalence was the expensive endotherm cellular toolkit: more muscle mitochondria and their associated proton leakage. Roger used this argument to emphasize that, unless large crocodile-like dinosaurs had been endotherms, they would not have been able to compete with similar-sized or even smaller endothermic Jurassic and Cretaceous mammals.

But the story of endothermy in lizards has just become more intriguing. Glenn Tattersall from Brock University in Ontario and his collaborators found evidence of endothermy in breeding tegu lizards, *Salvator merianae*.[8] At around 2 kg (4.4 lb), tegu lizards are comparatively big lizards. They occur in the tropics and semi-tropics of South America and fill the same ecological niche as monitor lizards do on other continents. After emerging from winter hibernation, tegu lizards increased their body temperatures by as much as 10°C (18°F) above the air temperature under controlled conditions. Apart from the shivering thermogenesis that has been reported in Burmese pythons, the Tattersall study is the first significant observation of internal heat production in a living reptile. The lizards increase their metabolic rates three-fold when they enter the breeding season. This increase in metabolic rate *does* lead to an increase in body temperature, unlike the paltry elevation seen in monitor lizards. What is going on here?

The authors do not offer any suggestion about where the heat is produced in the tegu lizard, but they do argue that the tegu is able to change its thermal conductance, most probably by dynamic constrictions and dilations of the vascular system, and thus conserve metabolic heat within its body. They argue that Brian McNab was right all along: as large lizards approach 10 kg (22.2 lb), their thermal conductance approaches those of similar-sized endothermic mammals. But why, they asked, has this never been observed before? Their answer: "That this degree of endothermy has not been reported in other similarly sized lizards (for example, varanid lizards) may be due to the use of techniques that promote peripheral blood flow and heat dissipation (that is, forced exercise enhances blood and air convection), thereby disrupting any endothermy bestowed by diminished thermal conductance."

As a generality, though, birds and mammals *must* possess body insulation if they are to elevate their body temperature through endothermy throughout the year. In modern mammals, which of course do have insulating fur, there is a strong positive relationship between basal metabolic rate and body temperature: the higher the basal metabolic rate, the higher the body temperature. I have called this the slow-fast metabolic continuum.[9]

Although the Aerobic Capacity Model was proposed specifically to explain the initial stages of the evolution of endothermy, there is no reason that its broad predictions should not be tethered to the later stages of endothermy as well. What is needed is a ubiquitous aerobic capacity model that can explain *all* pulses in aerobic capacity from the Carboniferous to the present, not just those that were associated with the earliest stages of the evolution of endothermy. We need a model that can explain how one hot little bird, the Arctic tern, can fly 81,000 km (50,331 mi) in one year and how the hot pronghorn can run so fast for so long.

For birds, though, we are presented with a slight dilemma; was the increased thigh muscle mass selected to produce heat on demand, for example to defend the body temperature, or was it selected, alternatively, to enhance activity? It is the thigh muscles, remember, that produce heat over and above that produced by the internal organs responsible for the basal metabolic rate. Both thermogenic capacity and cursoriality would have enhanced the dinosaur's fitness in various ways. This dichotomous scenario highlights the potential folly of peddling single-cause models to explain the evolution of endothermy rather than adopting approaches such as Kemp's Correlated Progression Model, which argues for multiple causes.

Despite the differences in muscle power that Roger Seymour has identified between crocodiles and mammals at the same temperature, I cannot bring myself to discard the role of body warmth as an important factor in sustained aerobic capacity. Perhaps one limitation of the Aerobic Capacity Model is that it downplays "warmth," possibly underestimating the importance of muscle temperature and *sustained* aerobic activity. Imagine this scenario. Let's suppose that the focus of selection was indeed for increased maximum metabolic rate that would allow feathered dinosaurs to utilize their ever-expanding heat-producing thigh muscles to run a bit faster. The Muscle Power Model would argue that the maximum efficiency of muscular performance requires that the muscles be warm on start-up and that this heat can be produced only by the internal organs.[10] Hence proportional selection for basal metabolic rate could have occurred concurrently in the internal organs to increase the muscle temperatures and hence facilitate higher maximum metabolic rates. Could this be a simple, multiple-cause explanation of the correlation between maximum and basal metabolic rates that does not require complex physiological or genetic linkages? We will not know the answer to this question until we start to factor body temperature into tests of the Aerobic Capacity Model.

One reason that we need an aerobic capacity model with a broader predictive scope than that proposed by Bennett and Ruben is that enhanced locomotor performance in mammals during the Cenozoic has been intricately tied up with supraendothermy, that is, the highest levels of endothermy ever attained in mammals. The pronghorn has the highest metabolic rate and body temperature of all mammals and also happens to be the mammal capable of the fastest sustained speed on earth. This is not a coincidence. It is the fastest mammals that have the highest metabolic rates and body temperatures. So I'm totally happy to argue that endothermy was selected for to increase activity in mammals. As part of my argument, though, I feel it essential to incorporate the Muscle Power Model into the mix. We need to better understand the intricacies of body warmth, not only how it is related to muscle power but, more importantly, how it is related to aerobic power.

The strongest challenge yet to the Aerobic Capacity Model has come from Colleen Farmer from the University of Utah, who proposed the Parental Care Model of endothermy in a paper published in the *American Naturalist* in 2000.[11] It is an odd paper because it reads like an angry rebuttal of the work of Michael Angilletta and Michael Sears on the costs of reproduction in lizards. In my opinion, though, it is the most likely and testable alternative to the Aerobic Capacity Model as to *why* endothermy evolved in birds and mammals initially. Farmer argues, quite simply, that temperature promotes the rate of development of young during incubation in birds and pregnancy in mammals. Here is a comment from Farmer's paper: "The evolution of parental provisioning indicates that minimizing foraging efforts, minimizing risks of adult predation, and minimization of daily energy expenditures were not favored over energy expenditure to reduce developmental time in the avian and mammalian lineages."[11] Pawel Koteja has added to the model by suggesting that an elevated metabolic rate also promotes post-hatching and postpartum care in the form of nestling food provision and lactation.[12]

I think that the Parental Care Model is useful because it has such tremendous and ubiquitous support not only from endotherms but also from ectotherms: sharks, lizards, and snakes. It also directly targets the most crucial time in an animal's life: when it breeds.

Several species of shark can maintain the uterus at an elevated temperature when pregnant by using a system of heat exchange that prevents heat from being lost to the peripheries of the animal and hence to seawater. However, the two best examples of heat production during reproduction in an ectothermic vertebrate are what have been observed in Burmese pythons, *Python bivittatus*, and in the tegu lizards I have discussed. Despite the fact that pythons and tegus belong to the Lepidosauria "ectothermic" lineage, they have the capacity to produce internal heat, by shivering or muscle twitching in Burmese pythons and probably by enhanced metabolism in the internal organs (and perhaps sarcolipin-SERCA muscle heat) in tegu lizards, but only when they brood their eggs. These lepidosaurs are able to increase their body temperatures significantly by 6–10°C (11–18°F) above the air temperature.[8,13] At an air temperature of 25°C (77°F) the metabolic rate of brooding snakes is nearly 100 times higher than that of nonbrooding snakes. Why would selection for a heat production mechanism in a large snake evolve and be expressed only when the snake is brooding were it not for the sole purpose of increasing the rate of development of egg-bound baby pythons?

It is worth clarifying a point here: a large ectotherm such as a python, contrary to the smaller force-fed lizards, can indeed increase its body temperature through shivering thermogenesis but probably not through the metabolic rate of its internal organs.

So what is the fitness benefit of speeding up embryo development?

There is no single, simple answer to this question. Its answers are embedded deep within modern life history theory, which relies on our understanding of trade-offs for explanations.

Let me try to provide a heuristic answer by using a hypothetical situation. Let's imagine a small mammal, such as a mouse, in two environments, one frequented by lots of predators and the other a rather safe, benign Eden sort of place where there are fewer predators, but predators nevertheless. The latter is not an unreal situation; it is typical of islands. Let's call these two mice the unsafe mouse and the safe mouse. The chance of the unsafe mouse surviving to breeding age is much lower than that of the safe mouse simply because its likelihood of being eaten by a predator is so much higher.

For the unsafe mouse, one way to overcome this problem is to breed earlier in life, which effectively reduces the time of its exposure to predators before breeding. For the safe mouse, its age at breeding is really not an issue; it may be determined by alternative factors such as food availability. But here is the trade-off: the unsafe mouse has the fitness benefit of breeding early, but it pays a fitness price (cost) because its offspring will be smaller and more vulnerable than the safe mouse's bigger babies. Small babies result simply because it takes time to grow big. If the parents are small, the babies will also be small. Herein, though, lies the fitness benefit for the safe mouse: bigger, more robust babies. They are bigger because the safe mouse mother had more time to grow big before she had her first litter. The fitness cost to the safe mouse is the increased risk of being eaten by a predator during the delayed time to breeding.

In continental environments where predators flourish in normal numbers unaffected by humans, there is a fitness benefit to speeding up development: it enables early breeding. This was the very characteristic that got our ancestors, the therapsids, through the Triassic. It is a ubiquitous fitness benefit in all animals. Indeed, the age at first reproduction has been argued to be the life history characteristic that affects an animal more powerfully than any other.[14] Nevertheless, every animal on earth falls somewhere along this age-at-first-reproduction trade-off continuum. Virtually every life history characteristic is involved in a trade-off continuum: litter size, the size of babies, the age at weaning, gestation time, lactation time, and so on. These continua are popularly known as "slow-fast" life history continua.

What is interesting about heat production in pythons is that it may have evolved as a mechanism of heat production long before heat production occurred in therapsids or the theropod dinosaurs. If it occurred in snakes, it may also have occurred in the ancestor of the snake-lizard lineage and hence also the ancestor of the bird lineage. In other words, shivering thermogenesis may have been inherited from this ancestor in the separate snake and bird lineages. And if so, the very earliest fitness benefit of endothermy, certainly in the bird lineage, was probably parental investment of heat into the development of young. Why this may be an important observation is that it may force us to separate the evolution of endothermy into various phases in both mammals and birds, which is the basis of the model that I will present shortly.

Another reason that I like the Parental Care Model is that it explains why tenrecs are so weird. The full picture of how tenrecs hibernate and breed was not fully appreciated when any of the models for the evolution of endothermy were first proposed. If it was, I'm not convinced that so much time, effort, and money would have been thrown at testing the Aerobic Capacity Model so relentlessly over the past few decades. Tenrecs (and many small marsupials) do not need to be warm or to have a high aerobic capacity to be active. Tenrecs, especially *Setifer setosus*, for which Danielle collected such intriguing data, express a pattern of endothermy that is exactly what the Parental Care Model predicts. The females increase their metabolic rates and make an attempt to maintain a constant body temperature only when they breed, when they are pregnant, when they are lactating, and when they are both pregnant and lactating.

I have suggested that endothermy evolved in three phases in iterative increments (as diagrammed in Figure 1.1).[15] The first phase occurred during the Permian in synapsids and during the Permian and Triassic in sauropsids. The second phase occurred during the Triassic, Jurassic, and Cretaceous in synapsids and during the Triassic and Early Jurassic in sauropsids. The third phase occurred during the Cenozoic in synapsids and during the Cretaceous and the Cenozoic in sauropsids. This triphasic model points to parental care as the primary driving force in the first phase, but it cannot be considered to be mutually exclusive of other driving forces, especially selection for aerobic capacity. Parental care had everything to do with egg development and also with the provisioning of young with food in both the archaic mammals, the therapsids, and in the very first bipedal dinosaurs with hot thighs.[12] In the early stages of development in birds and mammals, large body size and the importance of homeothermy through high thermal inertia was the common condition.

I can imagine that selection for increased leakiness of the membranes of the internal organs (described in Appendix 1) and perhaps an increase in mitochondrial density might have occurred to some limited extent in both lineages during their big-body-size phase to elevate their body temperatures above the air temperature. Homeothermy would have been propped up by thermal inertia and perhaps by shivering thermogenesis during the breeding season. But the basal metabolic rates that these early homeotherms may have enjoyed would certainly have been much lower than what is seen today in birds and mammals of equivalent size. The second and the third pulses in endothermy were needed to attain the modern endothermic status.

Phase 1 was also the "land-conquering" phase for both synapsids and sauropsids. It was during this phase that these animals emerged from the shrinking swamps of the Late Carboniferous and Early Permian to conquer dry land. I discussed the innovations required to achieve this successfully in the opening chapters, but essentially they involved more efficient food processing and water conservation; an enlargement of the dentary bone and the migration of the anterior jawbones to the middle ear; an increased diversification of the teeth (heterodonty); a well-vascularized fibrolamellar bone; a reduction of the lumbar ribs, indicative of diaphragmatic costal breathing; a shift from a sprawling to an upright posture; the development of the secondary palate separating the nasal and buccal cavities; and the development of water-conserving nasal turbinates. Clearly these innovations were not confined to ways of increasing the rate of embryo development. So, parental care cannot have been the only driving force during Phase 1.

All of these innovations, in some way or another, indicate an increased demand for oxygen, that is, an increase in the capacity for aerobic metabolism. I'll return to the aerobic capacity angle shortly.

James Hopson from the University of Chicago has proposed another single-cause model for the evolution of endothermy during Phase 1 that, he claims, "is preferable to the 'parental care' model of Farmer and the 'correlated progression' model of Kemp for understanding the origin of mammalian endothermy."[16]

Here again we have a single-cause model trying to outdo other single-cause models and dismissing the possibility that several selection forces can, or did, operate in tandem. Nevertheless, Hopson's idea is catchy. He argues, quite correctly, as I discussed in the opening chapters, that the stem synapsids were "sit-and-wait" foragers. This means that they lay around, probably half-submerged in the swamps, waiting for some edible victim to get within striking distance. Then they would explode into a burst of activity to catch the victim and rely almost exclusively on anaerobic metabolism to achieve this very fast-moving session of activity. They would then have needed to repay the oxygen debt later. This form of credit foraging still exists, for example, in ectothermic crocodiles.

However, sit-and-wait foraging is quite useless for an animal that wishes to colonize dry land successfully and competitively. Land colonizers needed to be able to do push-ups and to walk and run, which cost energy, and they needed to go out and find their food in order to meet the increasing energy demands of being more active. It was a vicious little energetic circle. Hopson calls land colonizers "widely foraging" feeders. Such animals cannot rely upon short bursts of anaerobic metabolism to do this.

In addition to what we can see and measure in the fossils of synapsids, such as the changes to the teeth, skulls, and post-cranial skeletons, there would have been soft-tissue innovations that did not fossilize but that must have arisen to allow aerobic capacity to occur. The entire cardiovascular and respiratory system would have shown improvements to enable more oxygen to be delivered to ever-demanding muscles and internal organs. Selection for increased aerobic capacity would therefore have occurred in tandem with that for parental care during Phase 1 and even thereafter.

I met Roger Seymour at the Ninth International Congress of Comparative Physiology and Biochemistry in Krakow, Poland, in August 2015. We chatted—not enough, sadly—over a beer about his presentation at the congress, in which he argued that the existence of a four-chambered heart and a one-way flow of air through the lungs of crocodiles suggests that endothermy must have existed in the ancestor of the crocodiles.[17]

A four-chambered heart is essential for generating a systemic blood pressure—that is, the blood pressure throughout the body—at levels high enough to sustain the elevated metabolic rates and gas exchanges needed for effective endothermy. In his presentation, Roger showed us data illustrating the massive difference in the blood pressures of noncrocodilian ectotherms and endotherms. He explained that, if the dinosaurs inherited a four-chambered heart from the ancestor that also gave rise to the crocodilian lineage, they would then have been able to sustain a blood pressure not much different from that of modern birds and mammals. In other words, he maintains that dinosaurs had the cardiovascular machinery for endothermy.

So crocs have thrown us a curveball. Roger argues that they may be the only group of animals that have reverted from an endothermic ancestry *back* to ectothermy. They reverted from the "widely-foraging" mode of the early land-dwelling crocodiles to the "sit-and-wait" foraging mode of crocs in aquatic habitats. As I have discussed, sit-and-wait foraging does not demand endothermy; it demands merely a reliance upon short-term anaerobic activity.

Crocodiles were the largest animals to survive the Cretaceous-Cenozoic extinction event, probably because of the protection they were afforded by water from the infrared thermal blast in the first hectic hours following the meteorite's impact. In the months that followed, they would have flourished on roast dinosaur and other cooked things. Their ectothermy would certainly have made their food go a long way for a long time.

Let us try to peg a date to the timing of the onset of Phase 1.

In the sauropsid lineage, Roger Seymour argues for a date associated with the evolution of the archosauriforms, the ancestors of the crocodile and dinosaur lineages—for example, stem archosaurs such as *Euparkeria*. His estimates are based upon the ontogeny of the development of the reptilian heart. Ontogeny is the development of an individual from the time of fertilization to maturity. The ontogeny of human development, for example, shows the existence of a tail in the human embryo, but by the time the baby is born, it has disappeared. The embryonic tail reflects the past evolutionary history of humans, namely, the fact that they evolved from mammals that did have tails. So Roger's estimates are based upon a tale of the reptilian heart as told by modern reptilian embryos. The heart evidence in the fossil record is, incidentally, not convincingly chronicled.

A highly promising but seemingly disregarded approach that Roger Seymour has been trying to emphasize for more than forty years and that *is* etched into the fossil record is the evidence that dinosaurs had high blood pressures, and were therefore endothermic.[18] With several colleagues, he also argued that the fossilized bones of synapsids and sauropsids leave tell-tale traces of high metabolic rates: in their femurs they have large foramina, holes in the bones whose diameters reflect the sizes of the arteries entering the bones, which are directly proportional to the amount of blood needed to fuel bone remodeling following activity.[17,19] This approach of analyzing the foramen size in the long bones holds enormous promise given the modern tools of phylogenetic reconstruction.

In Chapter 5 I argued that the Triassic reptiles *Euparkeria* and *Sclero-mochlus* might have been the first in the bird lineage to show signs of en-dothermy. I argued this on the basis of their being the first sauropsids to show signs of cursoriality in the form of long hind legs and short arms. This argument is based upon evidence, fossilized bones, so it augments very well arguments based upon the ontogeny of heart tissue development.

But *Scleromochlus taylori* was tiny, about the size of a starling, which does not fit with the idea that the first phase of endothermy occurred in large-bodied reptiles that enjoyed homeothermy through thermal inertia. How would such a small dinosaur have enjoyed the benefits of an elevated metabolic rate unless it was able to trap the metabolic heat produced with some form of insulation?

The problem in answering this question again revolves around the lack of hard evidence. Experts on the evolution of feathers argue that we should not exclude the possibility that feathers evolved much earlier in the avian lineage than the current data on fossilized feathers indicate. Walter Persons and Philip Currie from the University of Alberta have provided the most recent perspectives on the evolution of feathers. They argue:

> The current consensus regarding the phylogenetic distribution of
> endothermy among dinosaurs follows roughly the same pattern as
> the consensus regarding the distribution of feathers: endothermy,
> even if to only a primitive degree, is strongly affirmed throughout
> non avian coelurosaurs via multiple lines of evidence (perhaps
> the most compelling of which is the presence of feathers them-
> selves); endothermy as a trait of more ancestral theropod lineages

appears highly probable; endothermy as a basal trait of Dinosauria is a speculative notion, but not one without many discrete and tantalizing pieces of evidence in its favor [works of Robert Bakker cited], and the possibility that dinosaur endothermy actually originated at a point further down the archosaurian family tree is a hypothesis currently devoid of strong evidential support but nonetheless worthy of consideration and likely to be the subject of intense investigation in the coming years. Indeed, it is our own suspicion that the last hypothesis will ultimately prove correct.[20]

I would suggest that these authors consult Roger Seymour's 1976 publication on dinosaur blood pressures for "evidential support." The evidence and the ideas have been there all along, but they have been ignored.

The same problem of pinpointing the onset of increased metabolic rate applies to the ancestral mammals. Again we need to rely first on hard evidence—fossils—based upon remnants only, of what may really have existed. The argument for nasal turbinates' bones as the first evidence relies upon turbinal scars on the fossilized snouts of therapsids, thin lines etched into the sides of nasal bones. However, the data for *Lystrosaurus* in the Early Triassic showing an extensive network of cartilaginous maxilloturbinals, as well as fast growth rates, provide evidence that the thermal spike at the Permo-Triassic boundary may have accelerated the development of endothermy in the mammal lineage.

What is exciting, though, is the rapid emergence of the comparative method to estimate the metabolic status of archaic mammals and birds. This method simply marries measures of metabolic characteristics in modern animals and then uses the contemporary methods of phylogenetic reconstruction to map the characteristics back to the family tree to estimate the status of the ancestor at the base of the tree and of an extinct animal somewhere else deep in the tree.[21,22]

The timing of the onset of Phase 2 of endothermy is much easier to pinpoint, as are the main driving forces for additional elevations in the metabolic rates in the bird and mammal lineages. They are supported by good, solid fossil evidence: the first appearance of insulation (fur and feathers) and body-size miniaturization.

Brian McNab's hypothesis that miniaturization, for whatever reason drove it, was associated with strong selection for more heat per gram of tissue to compensate for size reduction,[23] is a neglected idea in the vocation of the evolution of endothermy. But it is an idea that now has tremendous support from fresh analyses showing a profound miniaturization in the bird lineage as well, albeit in two waves. Homeothermy, and hence its fitness benefits, could not have been retained with miniaturization unless both basal metabolic rate and heat production capacity were significantly enhanced.

We do not know what the driving forces were for body size miniaturization. In birds it was most certainly not driven by selection for flight capabilities, because the process commenced about 80 million years before the avian lineage took to the air. Indeed bone pneumatization commenced during the Late Triassic and was most evident in the very largest dinosaurs that roamed the earth. Lightweightness through bone pneumatization was selected for to counter gravity and to reduce the costs of locomotion on the ground. In addition, Roger Seymour argues that bone pneumatization allowed long-necked sauropods to float in their aquatic realms. In mammals, Brian McNab argued, miniaturization was driven by a change in resource availability as archaic mammals endured the Triassic. However, it was more probable that miniaturization is evident in the fossil record merely because the large-bodied therapsid and cynodont clades went extinct in the Early and Middle Triassic.[24] Moreover, small body size, fast growth rates, and early mortality seemed to be the crucial adaptations that allowed archaic mammals to survive the Early Triassic.[25]

I have called Phase 2 the "Thermoregulatory and Miniaturization Phase." I'd link the onset of Phase 2 to a pulse in endothermy driven by body size miniaturization and increased brain size in both birds and mammals. I'd also link its onset to the appearance of feathers and fur. Insulation in birds and mammals was essential for maintaining the heat within the body.

I call Phase 3 of endothermy the "Locomotory and Climate Adaptation Phase." In birds, muscle-powered flapping flight would have demanded an

increase in the basal metabolic rate and mitochondrial density and, arguably, muscle warmth, and it would have enabled higher rates of maximum
metabolic rate. If I am right, this phase would have preceded the onset of
the equivalent phase in mammals by more than 60 million years because
cursoriality and its enhanced metabolic demands evolved much later—in
the Eocene—in mammals. Selection for enhanced locomotion may also
have coincided with the parental care phase in large, feathered bipedal dinosaurs if the large thigh muscles were employed for running in addition to
thermogenesis.

It is quite fascinating that the three major characteristics that enabled
muscle-powered flapping flight in birds—increased metabolism, feathers,
and pneumatic, lightweight bones—all evolved long before the birds did.
All the right boxes had been ticked. Flapping flight was destined to happen.
All it needed for the final push was smaller body sizes, bigger brains, shorter
tails, bigger pectoral muscles, and higher metabolic rates, which, following
McNab's miniaturization idea in the latter case, were linked characteristics.

In both birds and mammals, the global cooling that occurred during
the Late Cenozoic had a profound influence on basal metabolic rates.
Endotherms that colonized the new cold high-latitude habitats needed to
have the heat-producing capacity to defeat the cold and defend their body
temperatures. High levels of heat production required elevated levels of
maximum metabolic rate, which, as we have seen, is linked to a high basal
metabolic rate. It is costly to inhabit the high latitudes.

The triphasic model of endothermy is obviously a multiple-causes model
that supports Kemp's Correlated Progression Model. Each successive endothermic phase would have built upon its predecessor in an iterative process, and indeed some overlap of driving evolutionary forces undoubtedly
occurred. For example, there is no reason to suppose that the benefits of
parental care during the second thermoregulation phase would have been
any less beneficial to fitness than they were in the first or last phases of
endothermy.

A phasic model of endothermy does not outright negate any of the former
models of the evolution of endothermy. Rather, the triphasic model is an

accommodating concatenation of published models pasted together with the aid of information from stunning Chinese fossils that have enabled us, for the first time, to piece together a reasonable timeline for the evolution of endothermy. The bird fossils, in particular, but also remarkable fossils of archaic mammals, have allowed us to pinpoint with more confidence the first appearance of features such as feathers and fur in birds and mammals.

The triphasic model invokes the Parental Care Model and the Aerobic Capacity Model as the driving forces for the first phase of endothermy. It invokes McNab's Miniaturization Model as the driving force for the second, thermoregulatory, phase of endothermy. But there is no current stand-alone model that can explain the third locomotory phase of endothermy.

The Aerobic Capacity Model, in the sense that it predicts selection for high maximum metabolic rates to sustain activity is conceptually equitable, but its denial of the importance of muscle temperature perhaps restricts its applicability to the very earliest stages of endothermy only. The Muscle Power Model provides a compelling physiological foundation for the locomotory phase but does not link warm muscles specifically to the evolution of new locomotor innovations.

In mammals, the Cenozoic Body Temperature Pulses Model and the Locomotory Modes Model do provide a timeline for the onset of the third phase.[26,27] These are my own models, and they state simply that the evolution of digitigrady (running on toes) and unguligrady (running on fingernails), which evolved in carnivores and ungulates, respectively, were associated with pulses in basal metabolic rate and body temperature during the Cenozoic. These pulses simply indicate increased aerobic capacity.

If the third locomotory phase in birds did indeed anticipate that in mammals by around 60 million years, it is probable that birds had already achieved their modern metabolic status long before the Cretaceous-Cenozoic boundary. The ancestral birds would have had a per-gram basal metabolic rate much higher than that of similar-sized mammals. I mention this because it makes it more difficult to explain how birds managed to survive the long-term resource-depleted period following the Chicxulub impact.

By virtue of its existence in modern poorwills, we cannot discount the possibility that hibernation was a characteristic of the avian lineage ancestral to modern birds. As I've argued for mammals, as the body sizes of feathered dinosaurs got smaller, there may have been a fitness benefit to toggling between ectothermy and endothermy on a daily basis, thus reducing predation risks, food demands, and the costs of staying warm. Those bird lineages that did get through the Cretaceous-Cenozoic boundary may have been capable of some degree of torpor or hibernation. Like crocodiles, birds may have had the advantage of protection from the initial infrared blast if they were periodically able to duck under the water. But this benefit would not have applied to the duck lineage (Anseriformes) only, because the other three lineages that survived the meteorite impact were land birds: the ancestral Palaeognathae (Cenozoic species including emus, cassowaries, moas, ostriches, rheas, kiwis, and elephant birds) and Galliformes (chickens and their kin). Moas and elephant birds are extinct, courtesy of humans, but the others are faring well.

The chicken lineage is the sister clade to the Anseriformes, and it is uncertain whether they split from the ducks before or after the Cretaceous-Cenozoic boundary.[28] If it was after the boundary, their survival is a nonissue because it is thought that the ancestor of the ducks and the chickens was a water bird. But if they split before the boundary, we need to figure out how these birds avoided getting fried, as for the Palaeognathae.

Perhaps the best explanation may be that they were able to avoid the infrared blast by seeking safe refuge. They would certainly have been small enough to hide among rocks, inside caves, or even in the burrows of other animals. Moreover, if the chickens in my garden are anything to go by, they were probably omnivorous and could eat just about anything, a great advantage in the lean months following the boundary.

I have no problem speculating endlessly about these possibilities because they pose evolutionary options that we may be able to test. The modern methods of character state reconstruction are marvelous tools for peeking into the past and for generating models of what ancestral birds and mammals were like, how they lived, and how they died.

Perhaps the most outstanding observation that can be made from my story of the evolution of endothermy is the remarkable synchrony of miniaturization in both the bird and the mammal lineages in the lead-up to full-blown thermoregulating endothermy by the Late Triassic. The same evolutionary forces did not necessarily drive miniaturization in both lineages, but they certainly produced the same outcome.

Living mammals are endotherms because their ancestors, the stem synapsids and therapsids, invaded dry land and managed to survive the thermal horrors of the Permo-Triassic boundary. For the rest of the Mesozoic, mammals managed to survive by growing fast, breeding early, dying young, and becoming nocturnal. Birds are endotherms because their ancestors also survived the Permo-Triassic nightmare, but, in addition, they went on to completely dominate the Triassic, Jurassic, and Cretaceous, mostly, it seems, during the day. Nevertheless, in many respects they too showed the survival life-history toolkit of the Triassic exhibited by the therapsids.

But thereafter the bird and mammal lineages had some luck. If the meteorite had not smashed into Earth in Mexico, mammals would not be as hot as they are today. They would still be small, nocturnal animals, secretive and illusive, yet as diverse a lineage as could be expected given their budding nocturnal endothermic capabilities under intense dinosaur dominance. We would have been at least 4°C (7.2°F) cooler than we are today, but warmish nevertheless. I doubt that men would have had scrotums.

Mammals owe their very existence and their hotness to a meteorite. They owe the meteorite for killing off the dinosaurs and the hugely diverse enantiornithine birds and opening up the earth for a new beginning, the age of the mammals. Placental mammals owe their existence to the persistence of Schrëwdinger through the dark, cold months that followed the impact because it had the capacity to abandon endothermy for months on end, just as do tenrecs, its close relatives, still do to this day on Madagascar.

Birds were the luckiest of all. They were full-blown insulated endotherms when the meteorite struck, yet they survived. All of their dinosaur relatives died out, many of which were also insulated endotherms. My guess is that it was their small body sizes and very nonspecialized diets that got them through because they avoided the infrared blast by hiding and perhaps hibernating. They could feed on anything that was on offer. So, in a way, they also owe their persistence to their colossal ornithischian, sauropodian, and theropodian cousins; the bigger their dinosaur relatives got, the more small niches were opened up for the smaller feathered dinosaurs to exploit.

As humans we owe our existence to the dramatic global cooling that commenced in the Late Eocene. From then until now, the planet cooled by an average of between 12° and 15°C (21–27°F), a staggering amount of heat that was sucked from the earth. Not only did the world cool steadily, but it also started to show massive bounces and oscillations in average temperatures over relatively short periods of geological time. These were the ice ages, and humans thrived on them; it caused the sizes of forests and intertidal zones to ebb and flow. The ice ages flushed the ancestors of humans out of the shrinking forests—especially during the cooler, drier phases—onto the open, grassy savannahs of Africa and onto the protein-rich shorelines. We are savannah and shoreline mammals. Our capacity for endothermy, our ability to defend our precious body temperatures—using the skins of other animals—and all of the fitness benefits that it guaranteed against the great temperature fluctuations of the ice ages—opened up the gates of Africa to the rest of the world: to Europe, Asia, Australasia, and finally North America. Humans became the most geographically widespread large-bodied species of mammal on earth, courtesy of the fires of life.

Heat on Demand

The minimum metabolic rate that is needed to keep an endotherm alive is called the basal metabolic rate. The metabolic machinery of the basal metabolic rate resides in the internal organs: the heart, lungs, kidneys, liver, and guts. In addition to this source of heat, endotherms need extra metabolic capacity to fuel activities such as running and flying, to process food, and to produce heat lost to a cold environment to defend the body temperature. If they cannot do this, they are not endotherms.

The physiological term for heat production is thermogenesis. Thermogenesis occurs in several tissue types in the body, and there are, according to Esa Hohtola from the University of Oulu, Finland, certain prerequisites that a tissue needs in order to generate heat in response to cold.[1] First, the tissue must occur in a sufficiently large tissue mass or be very intense in order to have enough capacity for defending the body against cooling. Second, the tissue must be under instantaneous nervous control so that the level of heat production can be controlled accurately in response to variations in the animal's thermal environment. And third, the tissue must be capable of heat production at all times of the year in order to be adaptive during seasonal changes in ambient temperature.

The extra heat over and above that which is produced by the basal metabolic rate comes in three forms: exercise-associated thermogenesis, non-exercise-associated thermogenesis, and diet-induced thermogenesis.

During a normal muscle fiber contraction, calcium (Ca^{2+}) is released from the lumen of the sarcoplasmic endoplasmic reticulum—the muscle version of the endoplasmic reticulum—via a calcium release channel (CRC) after the muscle fiber receives an action potential from the nervous system. The Ca^{2+} binds to the troponin component of the tropomyosin in the muscle fiber, causing it to contract, and in the process ATP is hydrolyzed in the myosin and heat is released. To complete the contraction cycle, sarcoplasmic endoplasmic reticulum Ca^{2+}-ATPase (SERCA) pumps the Ca^{2+} back into the lumen of the sarcoplasmic endoplasmic reticulum to restore the Ca^{2+} concentration. During this process, two Ca^{2+} molecules are pumped back and another ATP molecule is hydrolyzed: more heat. A third ATP molecule is hydrolyzed by Na$^+$-K$^+$-ATPase when the membrane potential is restored following the release of the calcium via the CRC: more heat. The heat generated in these muscle fiber contractions would be the heat that your muscles would produce, for example, when you went for a jog. It is the heat released when ATP is hydrolyzed during muscle contraction. This is exercise-associated thermogenesis.

Endotherms do not generally turn to exercise-associated thermogenesis if they are exposed to cold. You do not decide to go jogging when you start to feel cold. Nevertheless, exercise-associated thermogenesis can complement non-exercise-associated thermogenesis. If you do happen to be jogging in the cold, less non-exercise-associated thermogenesis will occur because of the heat produced by exercise-associated thermogenesis. So the more exercise thermogenesis that is produced, the less non-exercise-associated thermogenesis is required to defend the body temperature against the cold, roughly speaking.

Diet-induced thermogenesis occurs when you eat a meal. There is an increase in metabolism over and above the basal metabolic rate, and the source of this extra heat is the resynthesis of proteins. Diet-induced thermogenesis is influenced by what you eat and drink. Protein and alcohol induce the largest diet-induced thermogenic effects compared with carbohydrates and fats. Throughout a normal day, about 15 percent of your total energy requirements are tied up in diet-induced thermogenesis, more for oenophiles on a high-protein diet. But again, as in the case of exercise-associated thermogenesis, you do not start feasting on roast beef and drinking copious amounts of Cabernet Sauvignon when you feel the cold—except on rainy Sundays, of course. Diet-induced thermogenesis is not a form of heat that is relied upon to keep the body temperature constant.

The heat that we are most interested in from the perspective of the evolution of endothermy, though, is non-exercise-associated thermogenesis, and it comes in two forms: shivering thermogenesis and non-shivering thermogenesis.

Shivering thermogenesis is the process whereby muscle fibers undergo tremors or microvibrations without effecting any useful locomotor action or work. Muscles possess all of the important requisites for heat-producing tissues: they are the largest tissues in the body, they have an excellent nervous system control used normally for very precise muscle contraction, and they are active throughout the year. Despite a lack of effective action by the muscles, heat is nevertheless produced following the hydrolysis of ATP during shivering thermogenesis, as would occur normally during muscle contraction during activity. The heat is dissipated through the muscle and transported to the rest of the body via the circulation system. Shivering thermogenesis occurs in snakes (pythons), birds, and mammals. Shivering is, however, less efficient at producing heat than is non-shivering thermogenesis: "Only about one fifth of the input energy (in cellular fuels) is converted to external work, even in exercising muscles."[1]

The origins of shivering thermogenesis remain vague. Both birds and mammals use shivering thermogenesis, so perhaps the common ancestor of the synapsids and the sauropsids had the capacity for it and passed it on to their descendants. The alternative is that shivering thermogenesis evolved twice, independently, in the mammal and bird lineages. Shivering thermogenesis may therefore be one of the first forms of heat production on demand, controlled by the nervous system.

In mammals, non-shivering heat is produced in brown adipose tissue, so named because the fat is richly colored by the respiratory pigments of the electron transport chain in the mitochondria. Brown fat is densely packed with mitochondria. The brown fat mitochondria have specialized uncoupling proteins (UCP1) in their membranes that allow respiration to bypass or become uncoupled from the process of oxidative phosphorylation. Oxidative phosphorylation simply means the production of ATP from ADP using oxygen. This is the refueling process in the body, the regeneration of the body's gasoline, ATP.

During normal respiration, electrons from the Krebs cycle flow down the electron transport chain in the inner mitochondrial membrane and create a powerful proton-motive force on the outer surface of the inner mitochondrial membrane, that is, in the intermembrane space between the outer and inner mitochondrial membranes. The positively charged protons can't follow the negatively charged electrons across the membrane without access to some channel or other. The protons are blocked and accumulate on the outer surface of the inner mitochondria membrane. The normal channel across the membrane is the ATP synthase protein channel, which allows proton passage depending on the metabolic demand determined downstream from the ATP-synthase molecule. The demand downstream is governed simply by the amount of ADP floating around downstream. ADP is the rate-controlling molecule of respiration.

ADP is produced as soon as ATP is dephosphorylated or hydrolyzed during metabolic processes, for example, during muscle contractions. As ADP becomes available, protons flow through the ATP synthase and, with their accompanying electrons, are snapped up by oxygen to form water molecules. No oxygen, no proton flow. No ADP, no proton flow. So normal respiration is strictly controlled by the availability of both ADP and oxygen. Of course ADP is also phosphorylated to ATP, which is the whole purpose of respiration: the generation of biochemical fuel.

This strict control by ADP availability can be bypassed through UCP1 such that the ADP molecule is *not* phosphorylated to ATP but the protons still cross the mitochondrial membrane. The protons are snapped up by oxygen to form water molecules: metabolic water. As the protons flow through UCP1, they generate heat—a lot of heat—as the proton-motive force is dissipated. This is UCP1-mediated non-shivering thermogenesis, and its rate is now controlled by the sympathetic nervous system that activates brown adipose tissue.

Respiration is now controlled upstream from the inner membrane of the mitochondria, that is, on the supply side rather than the demand side. Whereas the supply of electrons normally comes from the Krebs cycle, for example, after the process of glycolysis—the breakdown of glucose—in non-shivering thermogenesis the electrons are generated during the breakdown of the brown fat in response to the nervous system. When fat is metabolized, it is broken down into its constituent components, triglycerides and glycerol. It is the triglycerides that are used to fuel non-shivering thermogenesis. The process is stimulated by the noradrenaline released from the sympathetic nervous system. Control now lies with the nervous system, so heat can be generated on demand in response, for example, to temperature-sensing neurons sending inputs to the hypothalamus signaling that more heat is needed for whatever reason.

Brown adipose tissue meets all of the criteria for a good heat-producing tissue. Although there is not a lot of brown adipose tissue in the body, compared with the amount of muscle tissue, its heat-generating capacity is phenomenal on a gram-for-gram basis. Brown adipose tissue is common in small mammals, such as rodents, and the neonates of large mammals, including infant humans. It is the most intense heat-producing tissue that exists in the animal kingdom. The non-shivering thermogenesis that the neonatal brown adipose tissue provides protects the baby from the cold in the absence, for example, of adequate parental care. Small mammals also employ UCP1-mediated non-shivering thermogenesis to warm up their bodies when they are being aroused from hibernation and daily torpor. But large mammals lose their brown adipose tissue as they mature because they switch to a reliance on heat produced by their skeletal muscles, of which neonates have very little.

Muscle can also engage in non-shivering thermogenesis because it can use ATP and generate heat without muscle fibers actually contracting. There is a protein in the membrane of the sarcoplasmic endoplasmic reticulum called sarcolipin that, when bound to SERCA, causes its function to become uncoupled from muscle fiber contraction. In essence, the Ca^{2+} that normally stimulates a muscle contraction is immediately pumped back into the sarcoplasmic reticulum by the SERCA-sarcolipin coupling without ever binding to troponin. So Ca^{2+} is cycled in a process that can go on for as long as there is neuronal stimulation in a continuous cycle of heat production. This is called a futile Ca^{2+} cycle.

Leslie Rowland and her colleagues from Ohio State University believe that once early birds and mammals started to increase the size of their skeletal muscles as they became more mobile when they conquered land, they had the potential to produce heat in the new muscles without necessarily using the muscles for locomotion.[2] They regard muscular non-shivering thermogenesis as the very first source of heat on demand, which appeared long before the advent of brown adipose tissue–derived non-shivering thermogenesis. Brown adipose tissue appeared for the first time in a primitive form in marsupials and evolved into its modern form in the placental mammals only. It does not occur in monotremes (echidnas and platypuses), nor does it occur in birds. Large mammals can rely on muscular non-shivering thermogenesis because they are, in general, cursors and therefore have large muscles.[2]

Birds do not have non-shivering heat production derived from UCP1. Interestingly, although the gene for UCP1 exists in all vertebrates (bony fish, amphibians, mammals), it got lost after the split of the vertebrate lineage into the mammal and bird lines.[3] It got lost somewhere in the ancestor of the lizards and the birds—lizards do not have the gene, either—during the long period of ectothermy maintained by the earliest reptiles. So birds do not have brown fat.

For typical ectothermic reptiles, such as lizards, snakes, and turtles, the loss of UCP1 was not a crisis. They did not need heat from brown fat, and so the loss of the gene was not noticed. But for other reptile lineages it was a crisis, according to Newman and his colleagues: "The direct ancestors of the theropod dinosaurs and the birds, was also among the group that lost UCP1. These animals, insofar as they were endothermic or even heterothermic, must have experienced the loss of the UCP1 gene as an existential crisis. This lineage was spared from extinction, according to the TMH [Thermogenic Muscle Hypothesis], only by selection for biochemical, physiological, and developmental novelties that facilitated enhanced thermogenesis and expansion of skeletal muscles."[4]

Again, with the remaining uncoupling protein genes at its disposal, natural selection found a way to produce heat from muscle, even without having to use the muscles for exercise. There are two other types of uncoupling protein, UCP2 and UCP3, and birds have lost UCP2 as well. But the bird lineage retained UCP3, which seems to be the protein that enables it to produce muscular heat, but not via proton leaks across the mitochondria membrane. How exactly UCP3 works is not clear, and the three possible hypotheses that exist at the time of this writing require more than high school biochemistry to understand. Let's just say that somehow UCP3 manages to produce heat within bird muscles without producing ATP.

It is fascinating that natural selection found ways to tinker with biochemical energy systems that had evolved and remained stable for hundreds of millions of years to generate novelties. Foremost is the concept of uncoupling in terms of heat production: the uncoupling of ATP generation in ATP-synthase by the uncoupling protein UCP1 in the non-shivering thermogenesis of brown adipose tissue and the uncoupling of muscle contraction from the SERCA pump during muscular non-shivering thermogenesis. ATP-synthase is one of the oldest biochemical molecules. It is found in all forms of life on earth, from yeast through viruses and from bacteria to wombats. It is truly ancient, so natural selection has had a long time to "look at it" and see how it can be exploited, bypassed, or accelerated in its normal function in respiration to ensure other functions.

Nasal Evaporative Cooling

The physics of nasal cooling is important in the evolution of endothermy. Consider a certain volume of air, let's say a cubic meter. This volume is composed of gases, each of which contributes a pressure, a partial pressure, to the total pressure that the volume of gas exerts in the atmosphere, along with all the other cubic meters of air. The common gases are oxygen, carbon dioxide, and nitrogen. Any given volume of air also holds water in the vapor phase, water vapor, which also has a partial pressure. However, whereas the partial pressures and amounts of oxygen, carbon dioxide, and nitrogen tend to remain the same over time in atmospheric air, the same is not true for water vapor. The amount of water vapor in the air varies considerably depending on factors such as proximity to standing water, latitude, season, and time of day. For example, on a typical day the air in tropical forests holds much more water vapor at noon than the air over the Atacama Desert.

When the cubic meter of air carries the maximum amount of water vapor that it can carry physically, it is fully saturated; it has a relative humidity of 100 percent. Any further water vapor added to this air will simply condense out on the nearest surface or as mist in the air. In the fully saturated state it contributes a maximum partial pressure of water vapor to the volume of air. When the air volume is not fully saturated, in other words, it has the physical capacity to take on more water vapor; then it is partially saturated, and the water vapor pressure is less. A volume of air with a relative humidity of 50 percent carries half of its total potential water vapor carrying capacity. The difference in the water vapor pressure between that at full saturation and that at partial saturation, at the same temperature, is called the *water vapor pressure saturation deficit,* and it controls the rate of evaporation from a wet surface. It is a hugely important measure in animal and plant physiology. But here's the crucial bit. Irrespective of the relative humidity of the air, be it 20 percent, 50 percent, or 100 percent, the partial pressure of water vapor in a given volume of air increases exponentially with temperature. The warmer the air, the more water vapor it can carry. This phenomenon has profound consequences for how animals balance a water account.

Consider a springbok on the grassy plains of the Kalahari Desert at midday. Say that the air temperature is 30°C (86°F) and the relative humidity is 20 percent. When the springbok inhales the dry, warm air it passes over the moist surfaces of the nasal passages and the breath of air heats up to body temperature, let's say to 38°C (100.4°F), and becomes fully saturated with water vapor in the lungs. The breath has picked up heat and water from the nasal regions, the trachea, and the lungs. If this breath of air were to be exhaled as is, warmed and fully saturated, the animal would lose a lot of water on each breath. There is nothing that the animal can do about the breath being exhaled fully saturated. There is no mechanism that I am aware of whereby a mammal can extract water and hence reduce the degree of saturation of exhaled air; ostriches and camels, though, can apparently expire unsaturated air.[1,2] If the temperature of the exhaled breath can be reduced, however, it will carry with it less water vapor, and in this way some of the water can be retrieved. To do this effectively, the nasal passages need to have a large surface area, and they need to be moist and cool, substantially cooler than the body temperature. Animals with large nostrils and a complex network of nasal turbinates can achieve this because evaporation takes place on inhalation, which cools the nasal surfaces. Evaporation is the most powerful mechanism for reducing the temperature of a moist surface.

Family Trees

Family trees, or phylogenies, are graphical representations that illustrate the relationships between species. From them it is possible, at a glance, to establish the closest relative(s) of a species and also to establish when that species originated and, if extinct, when it went extinct. In this appendix five family trees are provided (stem amphibians and the Synapsid-Sauropsid diversification, in Figure A.1; Therapsida, in Figure A.2; Mammaliaformes and Mammalia, in Figure A.3; Dinosauria, in Figure A.4; and Aves and Reptilia, in Figure A.5). In these five family trees every species or genus that is mentioned in the text of the book has been included. There remains considerable debate about whether the major radiations of the modern birds and placental mammals occurred prior to the Cretaceous-Cenozoic boundary (the long-fuse model) or immediately following the boundary (the short-fuse model). I have opted for the short-fuse models. The animal silhouettes were obtained from phylopic.org (public domain). The silhouettes are not shown to scale.

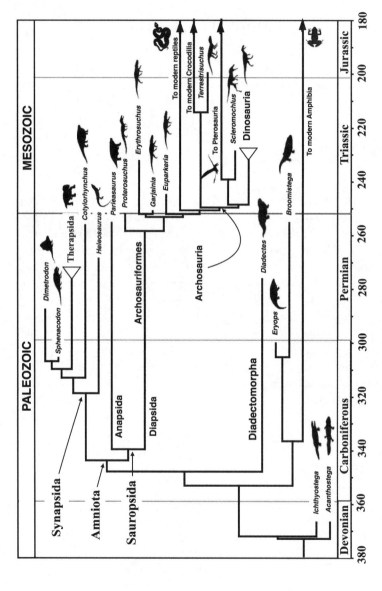

Figure A.1. The stem amphibians and the amniote diversification into the Sauropsida, which gave rise to the dinosaurs and the birds, and the Synapsida, which gave rise to the mammals. A triangle at the end of a lineage indicates that the clade is shown in a different figure (Therapsida, A.2, and Dinosauria, A.4). Divergence dates and phylogenies were sourced from the literature.[1-5] The numbers at the bottom of the figure represent millions of years ago.

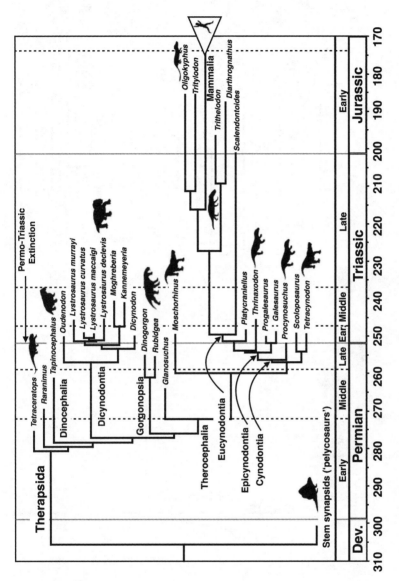

Figure A.2. Therapsida.[6-18] The triangle at the end of the Mammalia lineage indicates that the clade is shown in the next figure, Mammaliaformes and Mammalia. The numbers at the bottom of the figure represent millions of years ago.

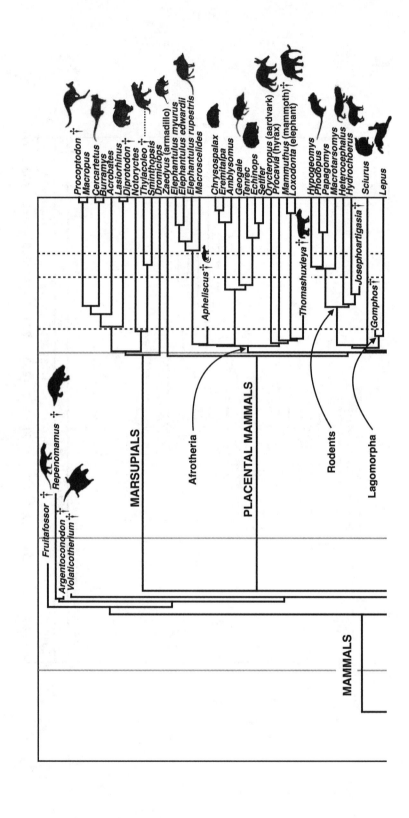

MAMMALS

MARSUPIALS

PLACENTAL MAMMALS

Afrotheria

Rodents

Lagomorpha

Fruitafossor †
Repenomamus †
Argentoconodon †
Volaticotherium †

Procoptodon †
Macropus
Cercartetus
Burramys
Acrobates
Lasiorhinus
Diprotodon †
Notoryctes †
Thylacoleo †
Sminthopsis
Dromiciops
Zaedyus (armadillo)
Elephantulus myurus
Elephantulus edwardii
Elephantulus rupestris
Macroscelides
Apheliscus †
Chrysospalax
Eremitalpa
Amblysomus
Geogale
Tenrec
Echinops
Setifer
Orycteropus (aardvark)
Procavia (hyrax)
Mammuthus (mammoth) †
Loxodonta (elephant)
Thomashuxleya †
Hypogeomys
Phodopus
Papagomys
Macrotarsomys
Heterocephalus
Hydrochoerus
Josephoartigasia †
Sciurus
Gomphos †
Lepus

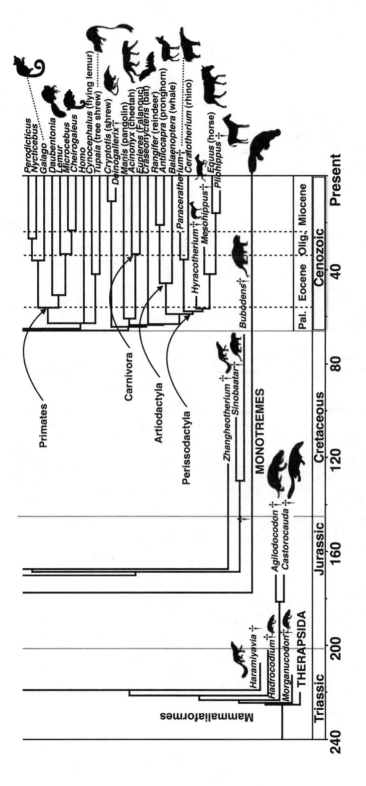

Figure A.3: Mammaliaformes and Mammalia.[19-35] Extinct species of mammaliaforms and mammals are shown with dagger symbols. The numbers at the bottom of the figure represent millions of years ago.

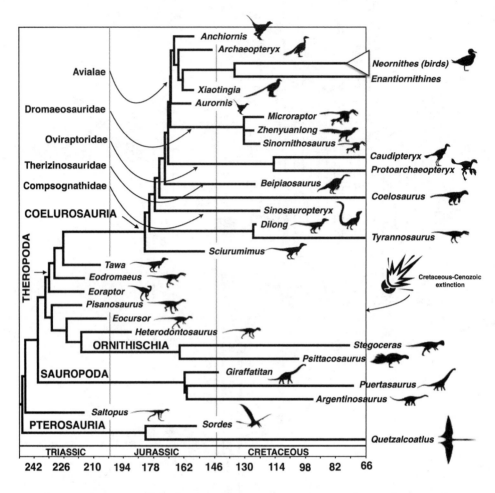

Figure A.4. Dinosauria.[36–40] The phylogenic arrangement of Sauropoda, Ornithischia, and Theropoda follows the arrangement of Baron et al.[36] All dinosaurs shown are extinct with the exception of the Aves (bird) clade, shown as a triangle, which means the clade is shown in the next figure, Aves and Reptilia. The numbers at the bottom of the figure represent millions of years ago.

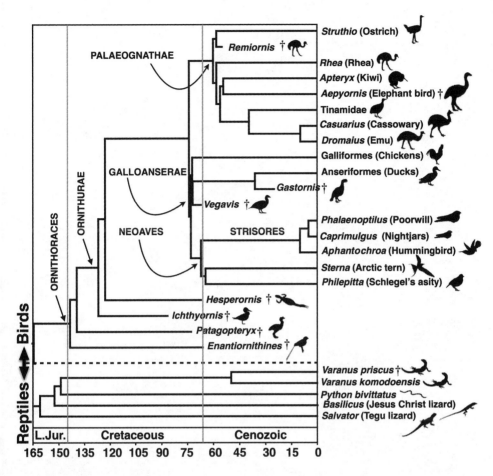

Figure A.5. Aves and Reptilia.[41–47] The phylogeny essentially follows Prum et al.,[42] but the age of the origination of the Galloanserae (Galliformes and Anseriformes) is shown as Late Cretaceous (rather than Early Cenozoic as in the phylogeny of Prum et al.), recognizing the fossil *Vegavis*, a Late Cretaceous basal duck (Anseriform) aged at about 67 million years old.[48] The dagger symbols indicate extinct species. The numbers at the bottom of the figure represent millions of years ago.

REFERENCES

Chapter 1. A Narrative of Travels

1. Wallace, A. R., *The Malay Archipelago* (New York: Cosimo, 2007).
2. Lovegrove, B. G., "The Evolution of Endothermy in Cenozoic Mammals: A Plesiomorphic-Apomorphic Continuum," *Biological Reviews* 87 (2012): 128–162.
3. Clarke, A., and Rothery, P., "Scaling of Body Temperature in Mammals and Birds," *Functional Ecology* 22 (2008): 58–67.

Chapter 2. Conquering Land

1. Schachat, S. R., et al., "Phanerozoic pO_2 and the Early Evolution of Terrestrial Animals," *Proceedings of the Royal Society B* 285: 20172631 (2018).
2. Harrison, J. F., Kaiser, A., and VandenBrooks, J. M., "Atmospheric Oxygen Level and the Evolution of Insect Body Size," *Proceedings of the Royal Society B* 277 (2010): 1937–1946.
3. Clapham, M. E., and Karr, J. A., "Environmental and Biotic Controls on the Evolutionary History of Insect Body Size," *Proceedings of the National Academy of Sciences of the United States of America* 109 (2012): 10927–10930.
4. Quammen, D., *The Song of the Dodo: Island Biogeography in an Age of Extinctions* (London: Pimlico, 1996).
5. Carroll, R. L., Irwin, J., and Green, D. M., "Thermal Physiology and the Origin of Terrestriality in Vertebrates," *Zoological Journal of the Linnean Society* 143 (2005): 345–358.
6. Clarke, A., and Pörtner, H.-O., "Temperature, Metabolic Power, and the Evolution of Endothermy," *Biological Reviews* 85 (2010): 703–727.
7. Ahlberg, P. E., Clack, J. A., and Blom, H., "The Axial Skeleton of the Devonian Tetrapod *Ichthyostega*," *Nature* 437 (2005): 137–140.

8. Pol, D., et al., "A New Fossil from the Jurassic of Patagonia Reveals the Early Basicranial Evolution and the Origins of Crocodyliformes," *Biological Reviews* 88 (2013): 862–872.

9. Pritchard, A. C., Turner, A. H., Allen, E. R., and Norell, M. A., "Osteology of a North American Goniopholidid (*Eutretauranosuchus delfsi*) and Palate Evolution in Neosuchia," *American Museum Novitates* (2013): 1–56.

10. Berman, D. S., et al., "Early Permian Bipedal Reptile," *Science* 290 (2000): 969–972.

11. Hungerbuehler, A., Mueller, B., Chatterjee, S., and Cunningham, D. P., "Cranial Anatomy of the Late Triassic Phytosaur *Machaeroprosopus*, with the Description of a New Species from West Texas," *Earth and Environmental Science Transactions of the Royal Society of Edinburgh* 103 (2012): 269–312.

12. Rayfield, E. J., Milner, A. C., Xuan, V. B., and Young, P. G., "Functional Morphology of Spinosaur 'Crocodile-Mimic' Dinosaurs," *Journal of Vertebrate Paleontology* 27 (2007): 892–901.

13. Sues, H.-D., Frey, E., Martill, D. M., and Scott, D. M., "*Irritator challengeri*, a Spinosaurid (Dinosauria: Theropoda) from the Lower Cretaceous of Brazil," *Journal of Vertebrate Paleontology* 22 (2002): 535–547.

14. Janis, C. M., and Keller, J. C., "Modes of Ventilation in Early Tetrapods: Costal Aspiration as a Key Feature of Amniotes," *Acta Palaeontologica Polonica* 46 (2001): 137–170.

15. Lillywhite, H. B., "Water Relations of Tetrapod Integument," *Journal of Experimental Biology* 209 (2006): 202–226.

16. Packard, M. J., and Seymour, R. S., "Evolution of the Amniote Egg," in *Amniote Origins*, ed. S. S. Sumida and K. L. M. Martin, 265–290 (London: Academic Press, 1997).

17. Modesto, S. P., and Anderson, J. S., "The Phylogenetic Definition of Reptilia," *Systematic Biology* 53, no. 5 (2004): 815–821.

Chapter 3. The Karoo

1. Broom, R., "A Further Contribution to Our Knowledge of the Fossil Reptiles of the Karroo," *Proceedings of the Zoological Society of London* 107 (1937): 299–318.

2. Rubidge, B., "The Roots of Early Mammals Lie in the Karoo: Robert Broom's Foundation and Subsequent Research Progress," *Transactions of the Royal Society of South Africa* 68 (2013): 41–52.

3. Liu, J., Rubidge, B., and Li, J. L., "New Basal Synapsid Supports Laurasian Origin for Therapsids," *Acta Palaeontologica Polonica* 54 (2009): 393–400.

4. Day, M. O., et al., "When and How Did the Terrestrial Mid-Permian Mass Extinction Occur? Evidence from the Tetrapod Record of the Karoo Basin, South Africa," *Proceedings of the Royal Society B* 282, 20150834 (2015).

5. Botha, J., and Smith, R. M. "*Lystrosaurus* Species Composition Across the Permo-Triassic Boundary in the Karoo Basin of South Africa," *Lethaia* 40 (2007): 125–137.

6. Broom, R. *On a New Reptile* (Proterosuchus fergusi) *from the Karoo Beds of Tarkastad, South Africa* (Trustees of the South African Museum and Geological Commission of the Colony of the Cape of Good Hope, 1903).

7. Botha-Brink, J., and Smith, R. M., "Osteohistology of the Triassic Archosauromorphs *Prolacerta, Proterosuchus, Euparkeria,* and *Erythrosuchus* from the Karoo Basin of South Africa, *Journal of Vertebrate Paleontology* 31 (2011): 1238–1254.

8. Gower, D. J., Hancox, P. J., Botha-Brink, J., Sennikov, A. G., and Butler, R. J., "A New Species of *Garjainia* Ochev, 1958 (Diapsida: Archosauriformes: Erythrosuchidae) from the Early Triassic of South Africa," *PLOS ONE* 9, e111154 (2014).

9. Broom, R., "Notice of Some New Fossil Reptiles from the Karoo Beds of South Africa," *Records of the Albany Museum* 1 (1905): 331–337.

10. Botha-Brink, J., Codron, D., Huttenlocker, A. K., Angielczyk, K. D., and Ruta, M., "Breeding Young as a Survival Strategy During Earth's Greatest Mass Extinction," *Scientific Reports* 6, 24053 (2016).

11. Smith, R. M., and Botha-Brink, J., "Anatomy of a Mass Extinction: Sedimentological and Taphonomic Evidence for Drought-Induced Die-Offs at the Permo-Triassic Boundary in the Main Karoo Basin, South Africa," *Palaeogeography, Palaeoclimatology, Palaeoecology* 396 (2014): 99–118.

12. Huttenlocker, A. K., and Botha-Brink, J., "Bone Microstructure and the Evolution of Growth Patterns in Permo-Triassic Therocephalians (Amniota, Therapsida) of South Africa," *PeerJ* 2, e325 (2014).

13. Huttenlocker, A. K., and Botha-Brink, J., "Body Size and Growth Patterns in the Therocephalian *Moschorhinus kitchingi* (Therapsida: Eutheriodontia) Before and After the End-Permian Extinction in South Africa," *Paleobiology* 39 (2013): 253–277.

14. Broom, R., "Note on *Mesosuchus browni*, Watson, and on a New South African Triassic Pseudosuchian (*Euparkeria capensis*)," *Records of the Albany Museum* 2 (1913): 394–396.

Chapter 4. The Permo-Triassic Kick-Start

1. Rey, K., et al., "Global Climate Perturbations During the Permo-Triassic Mass Extinctions Recorded by Continental Tetrapods from South Africa," *Gondwana Research* 37 (2016): 384–396.

2. Sun, Y., et al., "Lethally Hot Temperatures During the Early Triassic Greenhouse," *Science* 338 (2012): 366–370.

3. Botha-Brink, J., Codron, D., Huttenlocker, A. K., Angielczyk, K. D., and Ruta, M., "Breeding Young as a Survival Strategy During Earth's Greatest Mass Extinction," *Scientific Reports* 6, 24053 (2016).

4. Huttenlocker, A. K., and Farmer, C., "Bone Microvasculature Tracks Red Blood Cell Size Diminution in Triassic Mammal and Dinosaur Forerunners," *Current Biology* 27 (2016): 48–54.

5. Holland, R., and Forster, R., "The Effect of Size of Red Cells on the Kinetics of Their Oxygen Uptake," *Journal of General Physiology* 49 (1966): 727–742.

6. Rey, K., et al. "Oxygen Isotopes Suggest Elevated Thermometabolism Within Multiple Permo-Triassic Therapsid Clades," *eLife* 6, e28559 (2017).

7. Olivier, C., Houssaye, A., Jalil, N.-E., and Cubo, J., "First Palaeohistological Inference of Resting Metabolic Rate in an Extinct Synapsid, *Moghreberia nmachouensis* (Therapsida: Anomodontia)," *Biological Journal of the Linnean Society* 121 (2017): 409–419.

8. Poelmann, R. E., et al., "Evolution and Development of Ventricular Septation in the Amniote Heart," *PLOS ONE* 9, e106569 (2014).

9. Seymour, R. S., "Cardiovascular Physiology of Dinosaurs," *Physiology* 31 (2016): 430–441.

10. Smith, R. M., and Botha-Brink, J., "Anatomy of a Mass Extinction: Sedimentological and Taphonomic Evidence for Drought-Induced Die-Offs at the Permo-Triassic Boundary in the Main Karoo Basin, South Africa," *Palaeogeography, Palaeoclimatology, Palaeoecology* 396 (2014): 99–118.

11. Broom, R., "A Further Contribution to Our Knowledge of the Fossil Reptiles of the Karroo," *Proceedings of the Zoological Society of London* 107 (1937): 299–318.

12. Ruta, M., Botha-Brink, J., Mitchell, S. A., and Benton, M. J., "The Radiation of Cynodonts and the Ground Plan of Mammalian Morphological Diversity," *Proceedings of the Royal Society B* 280, 20131865 (2013).

13. Fernandez, V., et al., "Synchrotron Reveals Early Triassic Odd Couple: Injured Amphibian and Aestivating Therapsid Share Burrow," *PLOS ONE* 8, e64978 (2013).

14. Ruf, T., and Geiser, F., "Daily Torpor and Hibernation in Birds and Mammals," *Biological Reviews* 90 (2015): 891–926.

15. Broom, R., "On a New Type of Cynodont from the Stormberg," *Annals of the South African Museum* 7 (1913): 334–336.

Chapter 5. Reptile Takeover

1. Berman, D. S., et al., "Early Permian Bipedal Reptile," *Science* 290 (2000): 969–972.

2. Seymour, R. S., "Cardiovascular Physiology of Dinosaurs," *Physiology* 31 (2016): 430–441.

3. Seymour, R. S., Bennett-Stamper, C. L., Johnston, S. D., Carrier, D. R., and Grigg, G. C., "Evidence for Endothermic Ancestors of Crocodiles at the Stem of Archosaur Evolution," *Physiological and Biochemical Zoology* 77 (2004): 1051–1067.

4. Seymour, R. S., "Dinosaurs, Endothermy and Blood Pressure," *Nature* 262 (1976): 207–208.

5. Sander, P. M., et al., "Biology of the Sauropod Dinosaurs: The Evolution of Gigantism," *Biological Reviews* 86 (2011): 117–155.

6. Poelmann, R. E., et al., "Evolution and Development of Ventricular Septation in the Amniote Heart," *PLOS ONE* 9, e106569 (2014).

7. Crush, P. J., "A Late Upper Triassic Spenosuchid Crocodilian from Wales," *Palaeontology* 27 (1984): 131–157.

8. Crompton, A. W., Taylor, C. R., and Jagger, J. A. "Evolution of Homeothermy in Mammals," *Nature* 272 (1978): 333–336.

9. Lovegrove, B. G., and Mowoe, M. O., "The Evolution of Mammal Body Sizes: Responses to Cenozoic Climate Change in North American Mammals," *Journal of Evolutionary Biology* 26 (2013): 1317–1329.

10. Schmitz, L., and Motani, R., "Nocturnality in Dinosaurs Inferred from Scleral Ring and Orbit Morphology," *Science* 332 (2011): 705–708.

11. Heard-Booth, A. N., and Kirk, E. C., "The Influence of Maximum Running Speed on Eye Size: A Test of Leuckart's Law in Mammals," *Anatomical Record-Advances in Integrative Anatomy and Evolutionary Biology* 295 (2012): 1053–1062.

12. Clarke, A., and Pörtner, H.-O., "Temperature, Metabolic Power and the Evolution of Endothermy," *Biological Reviews* 85 (2010): 703–727.

13. Bakker, R. T., "Anatomical and Ecological Evidence of Endothermy in Dinosaurs," *Nature* 238 (1972): 81–85.

14. Newman, S. A., Mezentseva, N. V., and Badyaev, A. V., "Gene Loss, Thermogenesis, and the Origin of Birds," *Annals of the New York Academy of Sciences* 1289 (2013): 36–47.

Chapter 6. The Great Shrinking

1. McNab, B. K., "The Evolution of Endothermy in the Phylogeny of Mammals," *American Naturalist* 112, no. 983 (1978): 1–21.

2. Bakker, R. T., "Anatomical and Ecological Evidence of Endothermy in Dinosaurs," *Nature* 238 (1972): 81–85.

3. Luo, Z. X., Crompton, A. W., and Sun, A. L., "A New Mammaliaform from the Early Jurassic and Evolution of Mammalian Characteristics," *Science* 292 (2001): 1535–1540.

4. Harlow, H. J., Purwandana, D., Jessop, T. S., and Phillips, J. A., "Body Temperature and Thermoregulation of Komodo Dragons in the Field," *Journal of Thermal Biology* 35 (2010): 338–347.

5. Bennett, A. F., and Ruben, J. A., "Endothermy and Activity in Vertebrates," *Science* 206 (1979): 649–654.

6. Hillenius, W. J., "Turbinates in Therapsids: Evidence for Late Permian Origins of Mammalian Endothermy," *Evolution* 48 (1994): 207–229.

7. Bourke, J. M., et al., "Breathing Life into Dinosaurs: Tackling Challenges of Soft-Tissue Restoration and Nasal Airflow in Extinct Species," *Anatomical Record* 297 (2014): 2148–2186.

8. Laass, M., et al., "New Insights into the Respiration and Metabolic Physiology of *Lystrosaurus,*" *Acta Zoologica* 92 (2011): 363–371.

9. Lee, M. S., Cau, A., Naish, D., and Dyke, G. J., "Sustained Miniaturization and Anatomical Innovation in the Dinosaurian Ancestors of Birds," *Science* 345 (2014): 562–566.

10. Baron, M. G., Norman, D. B., and Barrett, P. M., "A New Hypothesis of Dinosaur Relationships and Early Dinosaur Evolution," *Nature* 543 (2017): 501–506.

11. Erickson, G. M., Rogers, K. C., Varricchio, D. J., Norell, M. A., and Xu, X., "Growth Patterns in Brooding Dinosaurs Reveal the Timing of Sexual Maturity in Non-Avian Dinosaurs and Genesis of the Avian Condition," *Biology Letters* 3 (2007): 558–561.

12. Huttenlocker, A. K., and Botha-Brink, J., "Body Size and Growth Patterns in the Therocephalian *Moschorhinus kitchingi* (Therapsida: Eutheriodontia) Before and After the End-Permian Extinction in South Africa," *Paleobiology* 39 (2013): 253–277.

13. Botha-Brink, J., Codron, D., Huttenlocker, A. K., Angielczyk, K. D., and Ruta, M., "Breeding Young as a Survival Strategy During Earth's Greatest Mass Extinction," *Scientific Reports* 6, 24053 (2016).

14. Smith, R. M., and Botha-Brink, J., "Anatomy of a Mass Extinction: Sedimentological and Taphonomic Evidence for Drought-Induced Die-Offs at the Permo-Triassic Boundary in the Main Karoo Basin, South Africa," *Palaeogeography, Palaeoclimatology, Palaeoecology* 396 (2014): 99–118.

Chapter 7. Feathers and Fur

1. Dhouailly, D., "A New Scenario for the Evolutionary Origin of Hair, Feather, and Avian Scales," *Journal of Anatomy* 214 (2009): 587–606.

2. Lillywhite, H. B., "Water Relations of Tetrapod Integument," *Journal of Experimental Biology* 209 (2006): 202–226.

3. McNab, B. K., "The Evolution of Endothermy in the Phylogeny of Mammals," *American Naturalist* 112, no. 983 (1978): 1–21.

4. Ji, Q., Luo, Z.-X., Yuan, C.-X., and Tabrum, A. R., "A Swimming Mammaliaform from the Middle Jurassic and Ecomorphological Diversification of Early Mammals," *Science* 311 (2006): 1123–1127.

5. Gower, D. J., Hancox, P. J., Botha-Brink, J., Sennikov, A. G., and Butler, R. J., "A New Species of *Garjainia* Ochev, 1958 (Diapsida: Archosauriformes: Erythrosuchidae) from the Early Triassic of South Africa," *PLOS ONE* 9, e111154 (2014).

6. Luo, Z.-X., Gatesy, S. M., Jenkins Jr., F. A., Amaral, W. W., and Shubin, N. H., "Mandibular and Dental Characteristics of Late Triassic Mammaliaform *Haramiyavia* and Their Ramifications for Basal Mammal Evolution," *Proceedings of the National Academy of Sciences* 112 (2015): E7101–E7109.

7. Lowe, C. B., et al., "Three Periods of Regulatory Innovation During Vertebrate Evolution," *Science* 333 (2011): 1019–1024.

8. Baron, M. G., Norman, D. B., and Barrett, P. M., "A New Hypothesis of Dinosaur Relationships and Early Dinosaur Evolution," *Nature* 543 (2017): 501–506.

9. Feduccia, A., *Riddle of the Feathered Dragons: Hidden Birds of China* (New Haven, CT: Yale University Press, 2014).

10. van der Reest, A. J., Wolfe, A. P., and Currie, P. J., "A Densely Feathered Ornithomimid (Dinosauria: Theropoda) from the Upper Cretaceous Dinosaur Park Formation, Alberta, Canada," *Cretaceous Research* 58 (2016): 108–117.

11. Lingham-Soliar, T., "The Evolution of the Feather: Scales on the Tail of *Sinosauropteryx* and an Interpretation of the Dinosaur's Opisthotonic Posture," *Journal of Ornithology* 154 (2013): 455–463.

12. Persons, W. S., and Currie, P. J., "Bristles Before Down: A New Perspective on the Functional Origin of Feathers," *Evolution* 69 (2015): 857–862.

13. Prum, R. O., and Brush, A. H., "The Evolutionary Origin and Diversification of Feathers," *Quarterly Review of Biology* 77 (2002): 261–295.

14. Prum, R. O., and Brush, A. H., "Which Came First, the Feather or the Bird?," *Scientific American* 23 (2014): 76–85.

15. Li, Q., et al., "Plumage Color Patterns of an Extinct Dinosaur," *Science* 327 (2010): 1369–1372.

16. Foth, C., "On the Identification of Feather Structures in Stem-Line Representatives of Birds: Evidence from Fossils and Actuopalaeontology," *Paläontologische Zeitschrift* 86 (2012): 91–102.

17. Norell, M. A., and Xu, X., "Feathered Dinosaurs," *Annual Review of Earth and Planetary Sciences* 33 (2005): 277–299.

18. Chen, P.-j., Dong, Z.-m., and Zhen, S.-n., "An Exceptionally Well-Preserved Theropod Dinosaur from the Yixian Formation of China," *Nature* 391 (1998): 147–152.

19. Qiang, J., Currie, P. J., Norell, M. A., and Shu-An, J., "Two Feathered Dinosaurs from Northeastern China," *Nature* 393 (1998): 753.

20. Rauhut, O. W., Foth, C., Tischlinger, H., and Norell, M. A., "Exceptionally Preserved Juvenile Megalosauroid Theropod Dinosaur with Filamentous Integument from the Late Jurassic of Germany," *Proceedings of the National Academy of Sciences* 109 (2012): 11746–11751.

21. Bell, P. R., et al., "Tyrannosauroid Integument Reveals Conflicting Patterns of Gigantism and Feather Evolution," *Biology Letters* 13, 20170092 (2017).

22. Amiot, R., et al., "Oxygen Isotopes of East Asian Dinosaurs Reveal Exceptionally Cold Early Cretaceous Climates," *Proceedings of the National Academy of Sciences* 108 (2011): 5179–5183.

23. Foth, C., Tischlinger, H., and Rauhut, O. W. M., "New Specimen of *Archaeopteryx* Provides Insights into the Evolution of Pennaceous Feathers," *Nature* 511 (2014): 79-U421.

24. Lowe, C. B., Clarke, J. A., Baker, A. J., Haussler, D., and Edwards, S. V., "Feather Development Genes and Associated Regulatory Innovation Predate the Origin of Dinosauria," *Molecular Biology and Evolution* 32 (2015): 23–28.

25. Lee, M. S., Cau, A., Naish, D., and Dyke, G. J., "Sustained Miniaturization and Anatomical Innovation in the Dinosaurian Ancestors of Birds," *Science* 345 (2014): 562–566.

26. Lü, J., and Brusatte, S. L., "A Large, Short-Armed, Winged Dromaeosaurid (Dinosauria: Theropoda) from the Early Cretaceous of China and Its Implications for Feather Evolution," *Scientific Reports* 5, 11775 (2015).

Chapter 8. Taking to the Air

1. Henderson, D. M., "Pterosaur Body Mass Estimates from Three-Dimensional Mathematical Slicing," *Journal of Vertebrate Paleontology* 30 (2010): 768–785.

2. Witton, M. P., and Habib, M. B., "On the Size and Flight Diversity of Giant Pterosaurs, the Use of Birds as Pterosaur Analogues and Comments on Pterosaur Flightlessness," *PLOS ONE* 5 (2010).

3. Marden, J. H., "From Damselflies to Pterosaurs: How Burst and Sustainable Flight Performance Scale with Size," *American Journal of Physiology* 266 (1994): R1077–R1084.

4. Seymour, R. S., "Cardiovascular Physiology of Dinosaurs," *Physiology* 31 (2016): 430–441.

5. Brusatte, S. L., Lloyd, G. T., Wang, S. C., and Norell, M. A., "Gradual Assembly of Avian Body Plan Culminated in Rapid Rates of Evolution Across the Dinosaur-Bird Transition," *Current Biology* 24 (2014): 2386–2392.

6. Benson, R. B., Butler, R. J., Carrano, M. T., and O'Connor, P. M., "Air-Filled Postcranial Bones in Theropod Dinosaurs: Physiological Implications and the 'Reptile'-Bird Transition," *Biological Reviews* 87 (2012): 168–193.

7. Xu, X., You, H., Du, K., and Han, F., "An *Archaeopteryx*-Like Theropod from China and the Origin of Avialae," *Nature* 475 (2011): 465–470.

8. Lee, M. S., and Worthy, T. H., "Likelihood Reinstates *Archaeopteryx* as a Primitive Bird," *Biology Letters* 8 (2012): 299–303.

9. Godefroit, P., et al., "A Jurassic Avialan Dinosaur from China Resolves the Early Phylogenetic History of Birds," *Nature* 498 (2013): 359–362.

10. Meredith, R. W., Zhang, G., Gilbert, M. T. P., Jarvis, E. D., and Springer, M. S., "Evidence for a Single Loss of Mineralized Teeth in the Common Avian Ancestor," *Science* 346, 1254390 (2014).

11. Dial, K. P., "Wing-Assisted Incline Running and the Evolution of Flight," *Science* 299 (2003): 402–404.

12. Balanoff, A. M., Bever, G. S., Rowe, T. B., and Norell, M. A., "Evolutionary Origins of the Avian Brain," *Nature* 501 (2013): 93–96.

13. Egevang, C., et al., "Tracking of Arctic Terns *Sterna paradisaea* Reveals Longest Animal Migration," *Proceedings of the National Academy of Sciences* 107 (2010): 2078–2081.

Chapter 9. The Heat of Darkness

1. Crompton, A. W., Taylor, C. R., and Jagger, J. A., "Evolution of Homeothermy in Mammals," *Nature* 272 (1978): 333–336.
2. Allman, J., "The Origin of the Neocortex," *Seminars in the Neurosciences* 2 (1990): 257–262.
3. Heesy, C. P., and Hall, M. I., "The Nocturnal Bottleneck and the Evolution of Mammalian Vision," *Brain Behavior and Evolution* 75 (2010): 195–203.
4. Benoit, J., Abdala, F., Van den Brandt, M. J., Manger, P. R., and Rubidge, B. S., "Physiological Implications of the Abnormal Absence of the Parietal Foramen in a Late Permian Cynodont (Therapsida)," *Science of Nature* 102 (2015): 1–4.
5. Gerkema, M. P., Davies, W. I., Foster, R. G., Menaker, M., and Hut, R. A., "The Nocturnal Bottleneck and the Evolution of Activity Patterns in Mammals," *Proceedings of the Royal Society B* 280, 20130508 (2013).
6. Oftedal, O. T., "The Mammary Gland and Its Origin During Synapsid Evolution," *Journal of Mammary Gland Biology and Neoplasia* 7 (2002): 225–252.
7. Kubo, T., and Kubo, M. O., "Nonplantigrade Foot Posture: A Constraint on Dinosaur Body Size," *PLOS ONE* 11, e0145716 (2016).
8. Grigg, G. C., Beard, L. A., and Augee, M. L., "The Evolution of Endothermy and Its Diversity in Mammals and Birds," *Physiological and Biochemical Zoology* 77 (2004): 982–997.
9. Kemp, T. S., "The Origin of Mammalian Endothermy: A Paradigm for the Evolution of Complex Biological Structure," *Zoological Journal of the Linnean Society* 147 (2006): 473–488.
10. Meng, J., "Mesozoic Mammals of China: Implications for Phylogeny and Early Evolution of Mammals," *National Science Review* 1 (2014): 521–542.
11. Rowe, T. B., Macrini, T. E., and Luo, Z. X., "Fossil Evidence on Origin of the Mammalian Brain," *Science* 332 (2011): 958–960.
12. Luo, Z. X., Crompton, A. W., and Sun, A. L., "A New Mammaliaform from the Early Jurassic and Evolution of Mammalian Characteristics," *Science* 292 (2001): 1535–1540.
13. Lovegrove, B. G., "The Low Basal Metabolic Rates of Marsupials: The Influence of Torpor and Zoogeography," in *Adaptations to the Cold: Tenth International Hibernation Symposium*, ed. F. Geiser, A. J. Hulbert, and S. C. Nicol, 141–151 (Armidale, Australia: University of New England Press, 1996).
14. Nicol, S. C., and Andersen, N. A., "Hibernation in the Echidna: Not an Adaptation to Cold?," in *Adaptations to the Cold: Tenth International Hibernation Symposium*, ed. F. Geiser, A. J. Hulbert, and S. C. Nicol, 7–12 (Armidale, Australia: University of New England Press, 1996).
15. Lovegrove, B. G., and McKechnie, A. E., *Hypometabolism in Animals: Hibernation, Torpor and Cryobiology* (Pietermaritzburg, South Africa: University of KwaZulu-Natal, 2008).

16. Luo, Z. X., "Transformation and Diversification in Early Mammal Evolution," *Nature* 450 (2007): 1011–1019.

17. Chen, M., and Wilson, G. P., "A Multivariate Approach to Infer Locomotor Modes in Mesozoic Mammals," *Paleobiology* 41 (2015): 280–312.

18. Hu, Y. M., Meng, J., Wang, Y. Q., and Li, C. K., "Large Mesozoic Mammals Fed on Young Dinosaurs," *Nature* 433 (2005): 149–152.

19. Gaetano, L. C., and Rougier, G. W., "New Materials of *Argentoconodon fariasorum* (Mammaliaformes, Triconodontidae) from the Jurassic of Argentina and Its Bearing on Triconodont Phylogeny," *Journal of Vertebrate Paleontology* 31 (2011): 829–843.

20. Wilson, G. P., et al., "Adaptive Radiation of Multituberculate Mammals Before the Extinction of Dinosaurs," *Nature* 483 (2012): 457–460.

Chapter 10. The Day of Reckoning

1. Alvarez, L. W., Alvarez, W., Asaro, F., and Michel, H. V., "Extraterrestrial Cause for the Cretaceous-Tertiary Extinction," *Science* 208 (1980): 1095–1108.

2. Robertson, D. S., McKenna, M. C., Toon, O. B., Hope, S., and Lillegraven, J. A., "Survival in the First Hours of the Cenozoic," *Geological Society of America Bulletin* 116 (2004): 760–768.

3. Kubo, T., and Kubo, M. O., "Nonplantigrade Foot Posture: A Constraint on Dinosaur Body Size," *PLOS ONE* 11, e0145716 (2016).

4. O'Leary, M. A., et al., "The Placental Mammal Ancestor and the Post-K-Pg Radiation of Placentals," *Science* 339 (2013): 662–667.

5. Clarke, J. A., Tambussi, C. P., Noriega, J. I., Erickson, G. M., and Ketcham, R. A., "Definitive Fossil Evidence for the Extant Avian Radiation in the Cretaceous," *Nature* 433 (2005): 305–308.

6. Jarvis, E. D., et al., "Whole-Genome Analyses Resolve Early Branches in the Tree of Life of Modern Birds," *Science* 346 (2014): 1320–1331.

7. Prum, R. O., et al., "A Comprehensive Phylogeny of Birds (Aves) Using Targeted Next-Generation DNA Sequencing," *Nature* 526 (2015): 569–U247.

8. McKechnie, A. E., and Lovegrove, B. G., "Avian Facultative Hypothermic Responses: A Review," *Condor* 104 (2002): 705–724.

9. Chinsamy, A., Chiappe, L. M., and Dodson, P., "Mesozoic Avian Bone Microstructure: Physiological Implications," *Paleobiology* 21 (1995): 561–574.

10. Geiser, F., "Evolution of Daily Torpor and Hibernation in Birds and Mammals: Importance of Body Size," *Clinical and Experimental Pharmacology and Physiology* 25 (1998): 736–740.

11. Riek, A., and Geiser, F., "Allometry of Thermal Variables in Mammals: Consequences of Body Size and Phylogeny," *Biological Reviews* 88 (2013): 564–572.

12. Jaeger, E. C., "Further Observations on the Hibernation of the Poor-Will," *Condor* (1949): 105–109.

13. McKechnie, A. E., Ashdown, R. A. M., Christian, M. B., and Brigham, R. M., "Torpor in an African Caprimulgid, the Freckled Nightjar *Caprimulgus tristigma*," *Journal of Avian Biology* 38 (2007): 261–266.

14. Smit, B., Boyles, J. G., Brigham, R. M., and McKechnie, A. E., "Torpor in Dark Times: Patterns of Heterothermy Are Associated with the Lunar Cycle in a Nocturnal Bird," *Journal of Biological Rhythms* 26 (2011): 241–248.

15. Morrow, G., and Nicol, S. C., "Cool Sex? Hibernation and Reproduction Overlap in the Echidna," *PLOS ONE* 4, e6070 (2009).

16. Geiser, F., "Hibernation and Daily Torpor in Two Pygmy Possums (*Cercartetus* Spp., Marsupialia)," *Physiological Zoology* 60 (1987): 93–102.

17. Geiser, F., "Yearlong Hibernation in a Marsupial Mammal," *Naturwissenschaften* 94 (2007): 941–944.

18. Ruf, T., and Geiser, F., "Daily Torpor and Hibernation in Birds and Mammals," *Biological Reviews* 90 (2015): 891–926.

19. Superina, M., and Boily, P., "Hibernation and Daily Torpor in an Armadillo, the Pichi (*Zaedyus pichiy*)," *Comparative Biochemistry and Physiology* A 148 (2007): 893–898.

20. Zack, S. P., Penkrot, T. A., Bloch, J. I., and Rose, K. D., "Affinities of 'Hyopsodontids' to Elephant Shrews and a Holarctic Origin of Afrotheria," *Nature* 434 (2005): 497–501.

Chapter 11. Clown Mouse and Golden Mole

1. Seguignes, M., "La Torpeur Chez *Elephantulus rozeti* (Insectivora: Macroscelididae)," *Mammalia* 47 (1983): 87–91.

2. Lovegrove, B. G., Lawes, M. J., and Roxburgh, L., "Confirmation of Pleisiomorphic Daily Torpor in Mammals: The Round-Eared Elephant Shrew *Macroscelides proboscideus* (Macroscelidea)," *Journal of Comparative Physiology* B 169 (1999): 453–460.

3. Lovegrove, B. G., Körtner, G., and Geiser, F., "The Energetic Cost of Arousal from Torpor in the Marsupial *Sminthopsis macroura*: Benefits of Summer Ambient Temperature Cycles," *Journal of Comparative Physiology* B 169 (1999): 11–18.

4. Geiser, F., et al., "Passive Rewarming from Torpor in Mammals and Birds: Energetic, Ecological and Evolutionary Implications," in *Life in the Cold: Evolution, Mechanisms, Adaptation, and Application*, ed. B. M. Barnes and H. V. Carey, 22–27 (Fairbanks: Institute of Arctic Biology, 2004).

5. Mzilikazi, N., and Lovegrove, B. G., "Daily Torpor During the Active Phase in Free-Ranging Rock Elephant Shrews (*Elephantulus myurus*)," *Journal of Zoology* 267 (2005): 103–111.

6. Boyles, J. G., Smit, B., Sole, C. L., and McKechnie, A. E., "Body Temperature Patterns in Two Syntopic Elephant Shrew Species During Winter," *Comparative Biochemistry and Physiology* A 161 (2012): 89–94.

7. Geiser, F., and Mzilikazi, N., "Does Torpor of Elephant Shrews Differ from That of Other Heterothermic Mammals?," *Journal of Mammalogy* 92 (2011): 452–459.

8. Ruf, T., and Geiser, F., "Daily Torpor and Hibernation in Birds and Mammals," *Biological Reviews* 90 (2015): 891–926.

9. Seymour, R. S., Withers, P. C., and Weathers, W. W., "Energetics of Burrowing, Running, and Free-Living in the Namib Desert Golden Mole (*Eremitalpa namibensis*)," *Journal of Zoology, London* 244 (1998): 107–117.

10. Fielden, L. J., Waggoner, J. P., Perrin, M. R., and Hickmann, G. C., "Thermoregulation in the Namib Desert Golden Mole, *Eremitalpa granti namibensis* (Chrysochloridae)," *Journal of Arid Environments* 18 (1990): 221–237.

11. Scantlebury, M., Lovegrove, B. G., Jackson, C. R., Bennett, N. C., and Lutermann, H., "Hibernation and Non-Shivering Thermogenesis in the Hottentot Golden Mole (*Amblysomus hottentottus longiceps*)," *Journal of Comparative Physiology B* 178 (2008): 887–897.

12. Lovegrove, B. G., and Wissel, C., "Sociality in Molerats: Metabolic Scaling and the Role of Risk Sensitivity," *Oecologia* 74 (1988): 600–606.

Chapter 12. Madness

1. Lovegrove, B. G., and Génin, F., "Torpor and Hibernation in a Basal Placental Mammal, the Lesser Hedgehog Tenrec *Echinops telfairi*," *Journal of Comparative Physiology B* 178 (2008): 691–698.

Chapter 13. Ankarafantsika

1. Choi, C. Q., "Did Mammals Sleep Through Cosmic Impact That Ended Dinosaurs?," *Live Science*, http://www.livescience.com/48389-did-mammals-hibernate-through-extinction.html (2014).

2. Quammen, D., *The Song of the Dodo: Island Biogeography and the Age of Extinctions* (London: Pimlico, 1996).

3. Goodman, S. M., Jungers, W. L., and Simeonovski, V., *Extinct Madagascar* (Chicago: University of Chicago Press, 2014).

4. Poux, C., Madsen, O., Glos, J., de Jong, W. W., and Vences, M., "Molecular Phylogeny and Divergence Times of Malagasy Tenrecs: Influence of Data Partitioning and Taxon Sampling on Dating Analyses," *BMC Evolutionary Biology* 8, no. 102 (2008): 1–6.

Chapter 14. Ancient Hibernation

1. Lovegrove, B. G., Lobban, K. D., and Levesque, D. L., "Mammal Survival at the Cretaceous-Paleogene Boundary: Metabolic Homeostasis in Prolonged Tropical Hibernation in Tenrecs," *Proceedings of the Royal Society B* 281 (2014).

2. Barnes, B., "Freeze Avoidance in a Mammal: Body Temperatures Below 0°C in an Arctic Hibernator," *Science* 244 (1989): 1593–1595.

3. Carey, H. V., Andrews, M. T., and Martin, S. L., "Mammalian Hibernation: Cellular and Molecular Responses to Depressed Metabolism and Low Temperature," *Physiological Reviews* 83 (2003): 1153–1181.

4. Fernandez, V., et al., "Synchrotron Reveals Early Triassic Odd Couple: Injured Amphibian and Aestivating Therapsid Share Burrow," *PLOS ONE* 8, e64978 (2013).

5. Poux, C., et al., "Asynchronous Colonization of Madagascar by the Four Endemic Clades of Primates, Tenrecs, Carnivores, and Rodents as Inferred from Nuclear Genes," *Systematic Biology* 54 (2005): 719–730.

6. Zhou, Y., Wang, S.-R., and Ma, J.-Z., "Comprehensive Species Set Revealing the Phylogeny and Biogeography of Feliformia (Mammalia, Carnivora) Based on Mitochondrial DNA," *PLOS ONE* 12, e0174902 (2017).

7. Simpson, G. G., "Mammals and Land Bridges," *Journal of the Washington Academy of Sciences* 30 (1940): 137–163.

8. Stankiewicz, J., Thiart, C., Masters, J., and Wit, M., "Did Lemurs Have Sweepstake Tickets? An Exploration of Simpson's Model for the Colonization of Madagascar by Mammals," *Journal of Biogeography* 33 (2006): 221–235.

9. Ali, J. R., and Huber, M., "Mammalian Biodiversity on Madagascar Controlled by Ocean Currents," *Nature* 463 (2010): 653–656.

10. Kappeler, P. M., "Lemur Origins: Rafting by Groups of Hibernators?," *Folia Primatologica* 71 (2000): 422–425.

11. Goodman, S. M., Jungers, W. L., and Simeonovski, V., *Extinct Madagascar* (Chicago: University of Chicago Press, 2014).

12. Masters, J. C., Lovegrove, B. G., and De Wit, M. J., "Eyes Wide Shut: Can Hypometabolism Really Explain the Primate Colonization of Madagascar?," *Journal of Biogeography* 34 (2007): 21–37.

13. Geiser, F. "Evolution of Daily Torpor and Hibernation in Birds and Mammals: Importance of Body Size," *Clinical and Experimental Pharmacology and Physiology* 25 (1998): 736–740.

14. Hildwein, G., and Goffart, M., "Standard Metabolism and Thermoregulation in a Prosimian, *Perodicticus potto*," *Comparative Biochemistry and Physiology* 50A (1975): 201–213.

15. Mzilikazi, N., Masters, J. C., and Lovegrove, B. G., "Lack of Torpor in Free-Ranging Southern Lesser Galagos, *Galago moholi*: Ecological and Physiological Considerations," *Folia Primatologica* 77 (2006): 465–476.

16. Nowack, J., Mzilikazi, N., and Dausmann, K. H., "Torpor on Demand: Heterothermy in the Non-Lemur Primate *Galago moholi*," *PLOS ONE* 5, e10797 (2010).

17. Ruf, T., Streicher, U., Stalder, G. L., Nadler, T., and Walzer, C., "Hibernation in the Pygmy Slow Loris (*Nycticebus pygmaeus*): Multiday Torpor in Primates Is Not Restricted to Madagascar," *Scientific Reports* 5, 17392 (2015).

18. Dausmann, K., "Flexible Patterns in Energy Savings: Heterothermy in Primates," *Journal of Zoology* 292 (2014): 101–111.

19. Ruf, T., and Geiser, F., "Daily Torpor and Hibernation in Birds and Mammals," *Biological Reviews* 90 (2015): 891–926.

20. Dewar, R. E., and Richard, A. F., "Evolution in the Hypervariable Environment of Madagascar," *Proceedings of the National Academy of Sciences, USA* 104 (2007): 13723–13727.

Chapter 15. Twenty-Four Nipples

1. Levesque, D. L., Lovasoa, O. M. A., Rakotoharimalala, S. N., and Lovegrove, B. G., "High Mortality and Annual Fecundity in a Free-Ranging Basal Placental Mammal, *Setifer setosus* (Tenrecidae: Afrosoricida)," *Journal of Zoology, London* 291 (2013): 205–212.

2. Farmer, C. G., "Parental Care: The Key to Understanding Endothermy and Other Convergent Features in Birds and Mammals," *American Naturalist* 155 (2000): 326–334.

3. Lovegrove, B. G., "Age at First Reproduction and Growth Rate Are Independent of Basal Metabolic Rate in Mammals," *Journal of Comparative Physiology B* 179 (2009): 391–401.

4. Levesque, D. L., and Lovegrove, B. G., "Increased Homeothermy During Reproduction in a Basal Placental Mammal," *Journal of Experimental Biology* 217 (2014): 1535–1542.

5. Stephenson, P. J., and Racey, P. A., "Reproductive Energetics of the Tenrecidae (Mammalia: Insectivora) I: The Large-Eared Tenrec, *Geogale aurita*," *Physiological Zoology* 66 (1993): 643–663.

6. Stephenson, P. J., "Reproductive Biology of the Large-Eared Tenrec, *Geogale aurita* (Insectivora: Tenrecidae)," *Mammalia* 57 (1993): 553–564.

Chapter 16. The Pronghorn Pinnacle

1. Alroy, J., "Cope's Rule and the Dynamics of Body Mass Evolution in North American Fossil Mammals," *Science* 280 (1998): 731–734.

2. Lovegrove, B. G., and Mowoe, M. O., "The Evolution of Mammal Body Sizes: Responses to Cenozoic Climate Change in North American Mammals," *Journal of Evolutionary Biology* 26 (2013): 1317–1329.

3. Demment, M. W., and Van Soest, P. J., "A Nutritional Explanation for Body-Size Patterns of Ruminant and Nonruminant Herbivores," *American Naturalist* 125 (1985): 641–672.

4. Dawson, T. J., Blaney, C. E., Munn, A. J., Krockenberger, A., and Maloney, S. K., "Thermoregulation by Kangaroos from Mesic and Arid Habitats: Influence of Tem-

perature on Routes of Heat Loss in Eastern Grey Kangaroos (*Macropus giganteus*) and Red Kangaroos (*Macropus rufus*)," *Physiological and Biochemical Zoology* 73 (2000): 374–381.

5. Wells, R. T., "Thermoregulation and Activity Rhythms in the Hairy-Nosed Wombat, *Lasiohinus latifrons* (Owen) (Vombatidae)," *Australian Journal of Zoology* 26 (1978): 639–651.

6. Flannery, T., *The Future Eaters* (New York: George Braziller, 1994).

7. Lovegrove, B. G., and Haines, L., "The Evolution of Placental Mammal Body Sizes: Evolutionary History, Form, and Function," *Oecologia* 138 (2004): 13–27.

8. Lovegrove, B. G., "The Zoogeography of Mammalian Basal Metabolic Rate," *American Naturalist* 156 (2000): 201–219.

9. Lovegrove, B. G., "Locomotor Mode, Maximum Running Speed and Basal Metabolic Rate in Placental Mammals," *Physiological and Biochemical Zoology* 77 (2004): 916–928.

10. Poux, C., Chevret, P., Huchon, D., de Jong, W. W., and Douzery, E. J. P., "Arrival and Diversification of Caviomorph Rodents and Platyrrhine Primates in South America," *Systematic Biology* 55 (2006): 228–244.

11. Rinderknecht, A., and Blanco, R. E., "The Largest Fossil Rodent," *Proceedings of the Royal Society B* 275 (2008): 923–928.

12. Simpson, G. G., *Splendid Isolation: The Curious History of South American Mammals* (New Haven, CT: Yale University Press, 1980).

13. Foster, J. B., "The Evolution of Mammals on Islands," *Nature* 202 (1964): 234–235.

14. Garland, T., "The Relation Between Maximal Running Speed and Body Mass in Terrestrial Mammals," *Journal of Zoology, London* 199 (1983): 157–170.

15. Lovegrove, B. G., and Mowoe, M. O., "The Evolution of Micro-Cursoriality in Mammals," *Journal of Experimental Biology* 217 (2014): 1316–1325.

16. White, C. R., Blackburn, T. M., and Seymour, R. S., "Phylogenetically Informed Analysis of the Allometry of Mammalian Basal Metabolic Rate Supports Neither Geometric nor Quarter-Power Scaling," *Evolution* 63 (2009): 2658–2667.

17. Lovegrove, B. G., "The Influence of Climate on the Basal Metabolic Rate of Small Mammals: A Slow-Fast Metabolic Continuum," *Journal of Comparative Physiology B* 173 (2003): 87–112.

18. Wiersma, P., Munoz-Garcia, A., Walker, A., and Williams, J. B., "Tropical Birds Have a Slow Pace of Life," *Proceedings of the National Academy of Sciences* 104 (2007): 9340–9345.

19. Jimenez, A. G., Cooper-Mullin, C., Calhoon, E. A., and Williams, J. B., "Physiological Underpinnings Associated with Differences in Pace of Life and Metabolic Rate in North Temperate and Neotropical Birds," *Journal of Comparative Physiology B* 184 (2014): 545–561.

Chapter 17. Cool Sperm

1. Lovegrove, B. G., "Cool Sperm: Why Some Placental Mammals Have a Scrotum," *Journal of Evolutionary Biology* 27 (2014): 801–814.

2. Sharma, V., Lehmann, T., Stuckas, H., Funke, L., and Hiller, M., "Loss of *RXFP2* and *INSL3* Genes in Afrotheria Shows That Testicular Descent Is the Ancestral Condition in Placental Mammals," *PLOS Biology* 16, e2005293 (2018): 1–22.

3. Frey, R., "On the Cause of the Mammalian Descent of the Testes (*Descensus Testiculorum*)," *Zeitschrift für Zoologische Systematik und Evolutionsforschung* 29 (1991): 40–65.

4. Gallup, G. G., Finn, M. M., and Sammis, B., "On the Origin of Descended Scrotal Testicles: The Activation Hypothesis," *Evolutionary Psychology* 7 (2009): 517–526.

5. Eagle, R. A., et al., "Body Temperatures of Modern and Extinct Vertebrates from ^{13}C-^{18}O Bond Abundances in Bioapatite," *Proceedings of the National Academy of Sciences* 107 (2010): 10377–11382.

6. Lovegrove, B. G., "The Evolution of Mammalian Body Temperature: The Cenozoic Supraendothermic Pulses," *Journal of Comparative Physiology B* 182 (2012): 579–589.

7. Short, R. V., "The Testis: The Witness of the Mating System, the Site of Mutation and the Engine of Desire," *Acta Paediatrica* 86 (1997): 3–7.

8. Bedford, J. M., "Enigmas of Mammalian Gamete Form and Function," *Biological Reviews* 79 (2004): 429–460.

9. Rommel, S. A., Pabst, D. A., McLellan, W. A., Mead, J. G., and Potter, C. W., "Anatomical Evidence for a Countercurrent Heat Exchanger Associated with Dolphin Testes," *Anatomical Record* 232 (1992): 150–156.

10. Clarke, A., and Pörtner, H.-O., "Temperature, Metabolic Power and the Evolution of Endothermy," *Biological Reviews* 85 (2010): 703–727.

11. Simpson, G. G., *Horses* (New York: Anchor, 1961).

12. Janis, C. M., and Wilhelm, P. D., "Were There Mammalian Pursuit Predators in the Tertiary? Dances with Wolf Avatars," *Journal of Mammalian Evolution* 1 (1993): 103–125.

13. Branch, M. P., "Ghosts Chasing Ghosts: Pronghorn and the Long Shadow of Evolution," *Ecotone* 4 (2008): 1–19.

14. Hebert, J., et al., "Thermoregulation in Pronghorn Antelope (*Antilocapra americana*, Ord) in Winter," *Journal of Experimental Biology* 211 (2008): 749–756.

15. Bennett, A. F., "The Evolution of Activity Capacity," *Journal of Experimental Biology* 160 (1991): 1–23.

16. Robergs, R. A., Ghiasvand, F., and Parker, D., "Biochemistry of Exercise-Induced Metabolic Acidosis," *American Journal of Physiology: Regulatory, Integrative and Comparative Physiology* 287 (2004): R502–R516.

17. Heard-Booth, A. N., and Kirk, E. C., "The Influence of Maximum Running Speed on Eye Size: A Test of Leuckart's Law in Mammals," *Anatomical Record: Advances in Integrative Anatomy and Evolutionary Biology* 295 (2012): 1053–1062.

18. Lovegrove, B. G., and Mowoe, M. O., "The Evolution of Micro-Cursoriality in Mammals," *Journal of Experimental Biology* 217 (2014): 1316–1325.

Chapter 18. Why We Are Hot

1. Bennett, A. F., and Ruben, J. A., "Endothermy and Activity in Vertebrates," *Science* 206 (1979): 649–654.
2. Angilletta, M. J., and Sears, M. W., "Is Parental Care the Key to Understanding Endothermy in Birds and Mammals?," *American Naturalist* 162 (2003): 821–825.
3. Swanson, D. L., "Are Summit Metabolism and Thermogenic Endurance Correlated in Winter-Acclimatized Passerine Birds?," *Journal of Comparative Physiology B* 171 (2001): 475–481.
4. Grossman, L. I., Wildman, D. E., Schmidt, T. R., and Goodman, M., "Accelerated Evolution of the Electron Transport Chain in Anthropoid Primates," *Trends in Genetics* 20 (2004): 578–585.
5. Nespolo, R. F., Bacigalupe, L. D., Figueroa, C. C., Koteja, P., and Opazo, J. C., "Using New Tools to Solve an Old Problem: The Evolution of Endothermy in Vertebrates," *Trends in Ecology and Evolution* 26 (2011): 414–423.
6. Bennett, A. F., Hicks, J. W., and Cullum, A. J., "An Experimental Test of the Thermoregulatory Hypothesis for the Evolution of Endothermy," *Evolution* 54 (2000): 1768–1773.
7. Seymour, R. S., "Maximal Aerobic and Anaerobic Power Generation in Large Crocodiles Versus Mammals: Implications for Dinosaur Gigantothermy," *PLOS ONE* 8, e69361 (2013).
8. Tattersall, G. J., Leite, C. A., Sanders, C. E., Cadena, V., Andrade, D. V., Abe, A. S., and Milsom, W. K., "Seasonal Reproductive Endothermy in Tegu Lizards," *Science Advances* 2, e1500951 (2016).
9. Lovegrove, B. G., "The Influence of Climate on the Basal Metabolic Rate of Small Mammals: A Slow-Fast Metabolic Continuum," *Journal of Comparative Physiology B* 173 (2003): 87–112.
10. Clarke, A., and Pörtner, H.-O., "Temperature, Metabolic Power and the Evolution of Endothermy," *Biological Reviews* 85 (2010): 703–727.
11. Farmer, C. G., "Parental Care: The Key to Understanding Endothermy and Other Convergent Features in Birds and Mammals," *American Naturalist* 155 (2000): 326–334.
12. Koteja, P., "Energy Assimilation, Parental Care and the Evolution of Endothermy," *Proceedings of the Royal Society of London Series B* 267 (2000): 479–484.
13. Brashears, J. A., and DeNardo, D. F., "Revisiting Python Thermogenesis: Brooding Burmese Pythons (*Python bivittatus*) Cue on Body, Not Clutch, Temperature," *Journal of Herpetology* 47 (2013): 440–444.
14. Stearns, S. C., *The Evolution of Life Histories* (Oxford: Oxford University Press, 1992).

15. Lovegrove, B. G., "A Phenology of the Evolution of Endothermy in Birds and Mammals," *Biological Reviews* 92 (2017): 1213–1240.

16. Hopson, J. A., "The Role of Foraging Mode in the Origin of Therapsids: Implications for the Origin of Mammalian Endothermy," *Fieldiana Life and Earth Sciences* (2012): 126–148.

17. Seymour, R. S., "Cardiovascular Physiology of Dinosaurs," *Physiology* 31 (2016): 430–441.

18. Seymour, R. S., "Dinosaurs, Endothermy, and Blood Pressure," *Nature* 262 (1976): 207–208.

19. Seymour, R. S., Smith, S. L., White, C. R., Henderson, D. M., and Schwarz-Wings, D., "Blood Flow to Long Bones Indicates Activity Metabolism in Mammals, Reptiles and Dinosaurs," *Proceedings of the Royal Society B* 279 (2012): 451–456.

20. Persons, W. S., and Currie, P. J., "Bristles Before Down: A New Perspective on the Functional Origin of Feathers," *Evolution* 69 (2015): 857–862.

21. Olivier, C., Houssaye, A., Jalil, N.-E., and Cubo, J., "First Palaeohistological Inference of Resting Metabolic Rate in an Extinct Synapsid, *Moghreberia nmachouensis* (Therapsida: Anomodontia)," *Biological Journal of the Linnean Society* 121 (2017): 409–419.

22. Huttenlocker, A. K., and Farmer, C., "Bone Microvasculature Tracks Red Blood Cell Size Diminution in Triassic Mammal and Dinosaur Forerunners," *Current Biology* 27 (2017): 48–54.

23. McNab, B. K., "The Evolution of Endothermy in the Phylogeny of Mammals," *American Naturalist* 112, no. 983 (1978): 1–21.

24. Sookias, R. B., Butler, R. J., and Benson, R. B., "Rise of Dinosaurs Reveals Major Body-Size Transitions Are Driven by Passive Processes of Trait Evolution," *Proceedings of the Royal Society B* (2012): rspb20112441.

25. Botha-Brink, J., Codron, D., Huttenlocker, A. K., Angielczyk, K. D., and Ruta, M., "Breeding Young as a Survival Strategy During Earth's Greatest Mass Extinction," *Scientific Reports* 6, 24053 (2016).

26. Lovegrove, B. G., "The Evolution of Mammalian Body Temperature: The Cenozoic Supraendothermic Pulses," *Journal of Comparative Physiology B* 182 (2012): 579–589.

27. Lovegrove, B. G., "Locomotor Mode, Maximum Running Speed and Basal Metabolic Rate in Placental Mammals," *Physiological and Biochemical Zoology* 77 (2004): 916–928.

28. Jarvis, E. D., et al., "Whole-Genome Analyses Resolve Early Branches in the Tree of Life of Modern Birds," *Science* 346 (2014): 1320–1331.

Appendix 1. Heat on Demand

1. Hohtola, E., "Shivering Thermogenesis in Birds and Mammals," in *Life in the Cold: Evolution, Mechanisms, Adaptation, and Application*, ed. B. M. Barnes and H. V.

Carey; paper presented at the Twelfth International Hibernation Symposium, Biological Papers of the University of Alaska, 27 (2004) (Fairbanks, Alaska: Institute of Arctic Biology, University of Alaska).

2. Rowland, L. A., Bal, N. C., and Periasamy, M., "The Role of Skeletal-Muscle-Based Thermogenic Mechanisms in Vertebrate Endothermy," *Biological Reviews* 90 (2014): 1279–1297.

3. Jastroch, M., Wuertz, S., Kloas, W., and Klingenspor, M., "Uncoupling Protein 1 in Fish Uncovers an Ancient Evolutionary History of Mammalian Nonshivering Thermogenesis," *Physiological Genomics* 22 (2005): 150–156.

4. Newman, S. A., Mezentseva, N. V., and Badyaev, A. V., "Gene Loss, Thermogenesis, and the Origin of Birds," *Annals of the New York Academy of Sciences* 1289 (2013): 36–47.

Appendix 2. Nasal Evaporative Cooling

1. Schmidt-Nielsen, K., *Animal Physiology: Adaptation and Environment* (Cambridge: Cambridge University Press, 1983).

2. Withers, P. C., *Comparative Animal Physiology* (Orlando: Saunders College, 1992).

Appendix 3. Family Trees

1. Swartz, B., "A Marine Stem-Tetrapod from the Devonian of Western North America," *PLOS ONE* 7, e33683 (2012).

2. Ruta, M., Pisani, D., Lloyd, G. T., and Benton, M. J., "A Supertree of Temnospondyli: Cladogenetic Patterns in the Most Species-Rich Group of Early Tetrapods," *Proceedings of the Royal Society B* 274 (2007): 3087–3095.

3. Benson, R. B. J., "Interrelationships of Basal Synapsids: Cranial and Postcranial Morphological Partitions Suggest Different Topologies," *Journal of Systematic Palaeontology* 10 (2012): 601–624.

4. Andres, B., Clark, J., and Xu, X. "The Earliest Pterodactyloid and the Origin of the Group," *Current Biology* 24 (2014): 1011–1016.

5. Nesbitt, S. J., "The Early Evolution of Archosaurs: Relationships and the Origin of Major Clades," *Bulletin of the American Museum of Natural History* 352 (2011): 1–288.

6. Liu, J., and Olsen, P., "The Phylogenetic Relationships of Eucynodontia (Amniota: Synapsida)," *Journal of Mammalian Evolution* 17 (2010): 151–176.

7. Day, M. O., Rubidge, B. S., and Abdala, F., "A New Mid-Permian Burnetiamorph Therapsid from the Main Karoo Basin of South Africa and a Phylogenetic Review of Burnetiamorpha," *Acta Palaeontologica Polonica* 61 (2016): 701–719.

8. Day, M. O., et al., "When and How Did the Terrestrial Mid-Permian Mass Extinction Occur? Evidence from the Tetrapod Record of the Karoo Basin, South Africa," *Proceedings of the Royal Society B* 282, 20150834 (2015).

9. Kemp, T. S., "The Origin and Early Radiation of the Therapsid Mammal-Like Reptiles: A Palaeobiological Hypothesis," *Journal of Evolutionary Biology* 19 (2006): 1231–1247.

10. Kammerer, C. F., "Systematics of the Rubidgeinae (Therapsida: Gorgonopsia)," *PeerJ* 4, e1608 (2016).

11. Huttenlocker, A., "An Investigation into the Cladistic Relationships and Monophyly of Therocephalian Therapsids (Amniota: Synapsida)," *Zoological Journal of the Linnean Society* 157 (2009): 865–891.

12. Abdala, F., Rubidge, B. S., and Van Den Heever, J., "The Oldest Therocephalians (Therapsida, Eutheriodontia) and the Early Diversification of Therapsida," *Palaeontology* 51 (2008): 1011–1024.

13. Rubidge, B. S., and Sidor, C. A., "Evolutionary Patterns Among Permo-Triassic Therapsids," *Annual Review of Ecology and Systematics* 32 (2001): 449–480.

14. Rubidge, B. S., and Hobson, J. A., "A New Anomodont Therapsid from South Africa and Its Bearing on the Ancestry of Dicynodontia," *South African Journal of Science* 86 (1990): 43–45.

15. Sigurdsen, T., Huttenlocker, A. K., Modesto, S. P., Rowe, T. B., and Damiani, R., "Reassessment of the Morphology and Paleobiology of the Therocephalian *Tetracynodon darti* (Therapsida), and the Phylogenetic Relationships of Baurioidea," *Journal of Vertebrate Paleontology* 32 (2012): 1113–1134.

16. Huttenlocker, A. K., Sidor, C. A., and Smith, R. M., "A New Specimen of *Promoschorhynchus* (Therapsida: Therocephalia: Akidnognathidae) from the Lower Triassic of South Africa and Its Implications for Theriodont Survivorship Across the Permo-Triassic Boundary," *Journal of Vertebrate Paleontology* 31 (2011): 405–421.

17. Botha-Brink, J., Codron, D., Huttenlocker, A. K., Angielczyk, K. D., and Ruta, M., "Breeding Young as a Survival Strategy During Earth's Greatest Mass Extinction," *Scientific Reports* 6, 24053 (2016).

18. Kammerer, C. F., Angielczyk, K. D., and Fröbisch, J., "A Comprehensive Taxonomic Revision of *Dicynodon* (Therapsida, Anomodontia) and Its Implications for Dicynodont Phylogeny, Biogeography, and Biostratigraphy," *Journal of Vertebrate Paleontology* 31 (2011): 1–158.

19. Cardillo, M., Bininda-Emonds, O. R. P., Boakes, E., and Purvis, A., "A Species-Level Phylogentic Supertree of Marsupials," *Journal of Zoology, London* 264 (2004): 11–31.

20. O'Leary, M. A., et al., "The Placental Mammal Ancestor and the Post-K-Pg Radiation of Placentals," *Science* 339 (2013): 662–667.

21. Herrera, J. P., and Dávalos, L. M., "Phylogeny and Divergence Times of Lemurs Inferred with Recent and Ancient Fossils in the Tree," *Systematic Biology* 65 (2016): 772–791.

22. Asher, R. J., et al., "Stem Lagomorpha and the Antiquity of Glires," *Science* 307 (2005): 1091–1094.

23. MacFadden, B. J., "Fossil Horses: Evidence for Evolution," *Science* 307 (2005): 1728–1730.

24. Fabre, P. H., Hautier, L., Dimitrov, D., and Douzery, E. J. P., "A Glimpse on the Pattern of Rodent Diversification: A Phylogenetic Approach," *BMC Evolutionary Biology* 12 (2012): 19.

25. Thiele, D., Razafimahatratra, E., and Hapke, A., "Discrepant Partitioning of Genetic Diversity in Mouse Lemurs and Dwarf Lemurs: Biological Reality or Taxonomic Bias?," *Molecular Phylogenetics and Evolution* 69 (2013): 593–609.

26. Zhou, Y., Wang, S.-R., and Ma, J.-Z., "Comprehensive Species Set Revealing the Phylogeny and Biogeography of Feliformia (Mammalia, Carnivora) Based on Mitochondrial DNA," *PLOS ONE* 12, e0174902 (2017).

27. Everson, K. M., Soarimalala, V., Goodman, S. M., and Olson, L. E., "Multiple Loci and Complete Taxonomic Sampling Resolve the Phylogeny and Biogeographic History of Tenrecs (Mammalia: Tenrecidae) and Reveal Higher Speciation Rates in Madagascar's Humid Forests," *Systematic Biology* 65 (2016): 890–909.

28. Douady, C. J., Catzeflis, F., Raman, J., Springer, M. S., and Stanhope, M. J., "The Sahara as a Vicariant Agent, and the Role of Miocene Climatic Events, in the Diversification of the Mammalian Order Macroscelidea (Elephant Shrews)," *Proceedings of the National Academy of Sciences* 100 (2003): 8325–8330.

29. Gillespie, A. K., Archer, M., and Hand, S. J., "A Tiny New Marsupial Lion (Marsupialia, Thylacoleonidae) from the Early Miocene of Australia," *Palaeontologia Electronica* 19 (2016): 1–25.

30. Black, K. H., Price, G. J., Archer, M., and Hand, S. J., "Bearing Up Well? Understanding the Past, Present, and Future of Australia's Koalas," *Gondwana Research* 25 (2014): 1186–1201.

31. Luo, Z.-X., Gatesy, S. M., Jenkins Jr., F. A., Amaral, W. W., and Shubin, N. H., "Mandibular and Dental Characteristics of Late Triassic Mammaliaform *Haramiyavia* and Their Ramifications for Basal Mammal Evolution," *Proceedings of the National Academy of Sciences* 112 (2015): E7101–E7109.

32. Bickelmann, C., et al., "The Molecular Origin and Evolution of Dim-Light Vision in Mammals," *Evolution* 69 (2015): 2995–3003.

33. Beck, R. M. D., Warburton, N. M., Archer, M., Hand, S. J., and Aplin, K. P., "Going Underground: Postcranial Morphology of the Early Miocene Marsupial Mole *Naraboryctes philcreaseri* and the Evolution of Fossoriality in Notoryctemorphians," *Memoirs of Museum Victoria* 74 (2016): 151–171.

34. Wilson, R. W., "Late Cretaceous (Fox Hills) Multituberculates from the Red Owl Local Fauna of Western South Dakota," *Dakoterra* 3 (1987): 118–122.

35. Bi, S., Wang, Y., Guan, J., Sheng, X., and Meng, J. "Three New Jurassic Euharamiyidan Species Reinforce Early Divergence of Mammals," *Nature* 514 (2014): 579–584.

36. Baron, M. G., Norman, D. B., and Barrett, P. M., "A New Hypothesis of Dinosaur Relationships and Early Dinosaur Evolution," *Nature* 543 (2017): 501–506.

37. Bates, K. T., et al., "Temporal and Phylogenetic Evolution of the Sauropod Dinosaur Body Plan," *Royal Society Open Science* 3, 150636 (2016).

38. Godefroit, P., et al., "A Jurassic Avialan Dinosaur from China Resolves the Early Phylogenetic History of Birds," *Nature* 498 (2013): 359–362.

39. Lü, J., and Brusatte, S. L., "A Large, Short-Armed, Winged Dromaeosaurid (Dinosauria: Theropoda) from the Early Cretaceous of China and Its Implications for Feather Evolution," *Scientific Reports* 5, 11775 (2015).

40. Hendrickx, C., Hartman, S. A., and Mateus, O., "An Overview of Non-Avian Theropod Discoveries and Classification," *PalArch's Journal of Vertebrate Palaeontology* 12 (2015): 1–73.

41. Jones, M. E., et al., "Integration of Molecules and New Fossils Supports a Triassic Origin for Lepidosauria (Lizards, Snakes, and Tuatara)," *BMC Evolutionary Biology* 13 (2013): 208.

42. Prum, R. O., et al., "A Comprehensive Phylogeny of Birds (Aves) Using Targeted Next-Generation DNA Sequencing," *Nature* 526 (2015): 569–U247.

43. Slack, K. E., et al., "Early Penguin Fossils, Plus Mitochondrial Genomes, Calibrate Avian Evolution," *Molecular Biology and Evolution* 23 (2006): 1144–1155.

44. Grealy, A., et al., "Eggshell Palaeogenomics: Palaeognath Evolutionary History Revealed Through Ancient Nuclear and Mitochondrial DNA from Madagascan Elephant Bird (*Aepyornis* Sp.) Eggshell," *Molecular Phylogenetics and Evolution* 109 (2017): 151–163.

45. White, N. D., Barrowclough, G. F., Groth, J. G., and Braun, M. J., "A Multi-Gene Estimate of Higher-Level Phylogenetic Relationships Among Nightjars (Aves: Caprimulgidae)," *Ornitología Neotropical* 27 (2016): 223–236.

46. Yonezawa, T., et al., "Phylogenomics and Morphology of Extinct Paleognaths Reveal the Origin and Evolution of the Ratites," *Current Biology* 27 (2017): 68–77.

47. Haddrath, O., and Baker, A. J., "Multiple Nuclear Genes and Retroposons Support Vicariance and Dispersal of the Palaeognaths, and an Early Cretaceous Origin of Modern Birds," *Proceedings of the Royal Society B*, rspb20121630 (2012).

48. Clarke, J. A., Tambussi, C. P., Noriega, J. I., Erickson, G. M., and Ketcham, R. A., "Definitive Fossil Evidence for the Extant Avian Radiation in the Cretaceous," *Nature* 433 (2005): 305–308.

INDEX

Page numbers in italics refer to illustrations

ATP (adenosine triphosphate), 22, 79, 216, 269, 279–280, 281, 287, 288, 307–308, 309, 310, 311
Augee, Michael, 128
Aurornis xui, 117
Australia: marsupials, 245–248; Pleistocene crocodiles and lizards, 71–72
Australopithicus, 46–47
Avialae, 116–117, 119–120. *See also* birds
aye-aye (*Daubentonia madagascariensis*), 224, 224–225

Bain, Andrew Geddes, 48
Bakker, Robert, xviii, 80, 83, 84
Balanoff, Amy, 119
Barnes, Brian, 213–215
Baron, Matthew, 91, 320
Barosaurus, 74
basal metabolic rate (BMR), 7, 257, 285–286, 289, 290, 301, 307
Basiliscus basiliscus (Jesus Christ lizard), 69
basoendotherms, 8, 9, 182, 233, 245, 258
Beard, Lyn, 128
Beipiaosaurus inexpectus, 106, 107–108
Bennett, Albert "Al," xviii, 87–88, 89, 279, 284–285, 287–288
Bennett, Nigel, 181–183
Benson, Roger, 115
Berenty tenrec study (Madagascar), 187–194, 233, 238–239
Berlin Museum of Natural History, 74–75, 116
Berman, David, 68–69
Bernard Price Institute (BPI), 42, 44
beta-keratin, 35, 95–96, 100–101, 103
bipedalism, 52, 68–69, 71, 74, 75, 81–82
birds: *Archaeopteryx* as, 115–117, 118, 119; bipedalism, 81–82; and Cretaceous-Cenozoic extinction event, 152–157, 302–303; effects of habitat/climate on, 259; evolution of flight in, 117–121; first evidence of endothermy in, 155–156, 298; and hibernation, 156–161; migratory, 120–121; miniaturization in, 90–93; phylogeny, 321; role of thigh muscle mass in,

290; spermatogenesis in, 268; and teeth, 117. *See also* sauropsid lineage
blood pressure, 59, 73–75, 296–297
BMR. *See* basal metabolic rate (BMR)
boas (Malagasy ground boas), 209, 231–232, 233–234, 239–240
body size: and blood pressure, 74; and body temperature, 245–247; changes in, and evolution of endothermy, 84–87; and digestion of plants, 244–245, 248–249; and evolution of angiosperms, 142; gigantism, 18–19, 253–254; and locomotor gaits and predation, 249–256; and mammals, during Cenozoic, 241; and metabolic rates, 256–259, 257; miniaturization (*see* miniaturization); predation and, 252–254; reconstruction of, for extinct mammals, 264; and speed, 282–283; surface-area-to-volume ratio (SA:V), 33–34, 34; and thermal conductance, 86–87
body temperature: and artificial selection, 276–277; and body size, 245–247; consequences of low body temperatures, 213; cursorial ungulates and carnivores, 276–282; estimating for extinct mammals, 57–59, 264–267; hypothermia, 5–6; as measure of endothermy, 7–9; set point, 7
Boily, Patrice, 166, 167
Bolosauridae, 32, 68
Bolt, Usain, 279, 280, 283
Botha-Brink, Jennifer, 49, 56, 60–61, 66, 93
Bourke, Jason, 89
Boyles, Justin, 178
Brachiosaurus (*Giraffatitan*), 74–75
brain size, 119, 123, 129–133
Branch, Michael, 274
breeding. *See* reproduction
Brigham, Mark, 157–158, 158
Broom, Robert, 15, 40, 42–44, 43, 45–46, 47, 50, 51, 52, 60, 65–66, 77, 89
Broomistega, 64, 64
brown adipose tissue, 309–311
brown lemurs, 204